Quantum Theory without Reduction

Quantum Theory without Reduction

Edited by

Marcello Cini

Università di Roma

and

Jean-Marc Lévy-Leblond

Université de Nice

Adam Hilger, Bristol and New York

© IOP Publishing Ltd 1990

British Library Cataloguing in Publication Data

Quantum theory without reduction.
  1. Quantum theory
  I. Cini, Marcello   II. Lévy-Leblond, Jean-Marc
  530.12

  ISBN 0-7503-0031-0

Library of Congress Cataloging-in-Publication Data are available

Published under the Adam Hilger imprint by IOP Publishing Ltd
Techno House, Redcliffe Way, Bristol BS1 6NX, England
335 East 45th Street, New York, NY 10017-3483, USA

Printed in Great Britain by Galliard (Printers) Ltd, Great Yarmouth, Norfolk

Contents

Preface

Quantum theory offers a strange, and perhaps unique, case in the history of science: it has a well-established body of knowledge, a huge bulk of experimental evidence, a growing amount of technical applications—while no consensus upon its conceptual basis has been reached. . . . Although research on the foundations of quantum theory has gained a new respectability and may have claimed some beautiful results in recent decades, the debate still goes on—in particular about the heated issue of measurement theory and, specifically, on the 'reduction of the state vector'.

Among various channels of thought, one seems to attract an increasing number of followers, despite the fact that it has not yet expressed itself in a consistent and explicit manner. We speak of those among us who assume the reduction of the state vector to be a false problem; our point of view is that quantum theory is a consistent and complete theory without this assumption, which, far from being a basic axiom, is an *ad hoc* recipe—admittedly a useful and working one—the validity of which may and must be established. Feeling that the time is ripe for a general presentation of this point of view, we have collected in this volume various specific contributions having the common goal of building what we feel to be a long overdue vision of quantum theory, which puts into a new light many of its long-standing problems.

As the reader will notice, here are different approaches to the same goal, namely, building a consistent understanding of quantum theory without assuming *a priori* the reduction postulate: studying the measuring apparatus as such by emphasizing its macroscopic nature; stressing the statistical aspects of a realistic analysis of the classical limit; proposing some specific and concrete models of measurements as a quantum physical process; or emphasizing the logical pitfalls of the orthodox views and building a new logical formulation of quantum theory.

We believe that the various papers gathered here converge to bring an old debate into a new perspective, and hope that they will stimulate further work, leading perhaps to a more satisfying consensus about quantum theory.

The basis for gathering and editing the present volume has been provided by an International Colloquium held at the University of Rome 'La Sapienza' in April 1989. The organization of this meeting would not have been possible without the generous financial support of the following institutions:

Università di Roma 'La Sapienza' e Facoltà di Scienze MFN dell'Università di Roma 'La Sapienza', Italy.
Instituto Nazionale di Fisica Nucleare, Italy.
Comitato per la Fisica, Consiglio Nazionale delle Ricerche, Italy.
Ministère des Affaires Etrangères; Direction de la Coopération Scientifique, Technique et du Développement, France.

January 1990

Marcello Cini Università di Roma
Jean-Marc Lévy-Leblond Université de Nice

List of Contributors

J-M Lévy-Leblond   Laboratoire de Physique Théorique, *Université de Nice*, Faculté des Sciences, Parc Valrose, 06034 Nice Cedex, France

A Zeilinger   *Atominstitut*, Schuettel Str. 115, A-1020 Wien, Austria

R Omnès   Laboratoire de Physique Théorique et Hautes Energies, *Université Paris-Sud*, Bat. 211, 91405 Orsay Cedex, France

H Primas   Laboratorium für Physikalisches Chemie, *ETH-Zentrum*, CH-8092 Zürich, Switzerland

R Fukuda   Dept. of Physics, Faculty of Science and Technology, *Keio University*, Hiyoshi 3–14–1, Yokohama 233, Japan

R Balian   Service de Physique Théorique, *C.E.A. Saclay*, 91191 Gif-sur-Yvette Cedex, France

M Cini   Dipartimento di Fisica, *Università di Roma 'La Sapienza'*, P.le A. Moro 2, 00185 Roma, Italy

M Serva   Dipartimento di Fisica, *Università di Roma 'La Sapienza'*, P.le A. Moro 2, 00185 Roma, Italy

A Peres   Department of Physics, *Israel Inst. of Technology Technion*, 32000 Haifa, Israel

Y Ben-Dov   The Institute for History and Philosophy of Science, *Tel Aviv University*, Ramat Aviv, 69978 Tel Aviv, Israel

E J Squires   Department of Mathematical Sciences, *University of Durham*, South Road, Durham, DH1 3LE, UK

P T Landsberg   Faculty of Mathematical Studies, *University of Southampton*, Southampton, SO9 5NH, UK

Where is the Problem?

Jean-Marc Lévy-Leblond

1. Introduction : *what* is the problem ?

The reduction of the state vector is rightly held as one of the most puzzling aspects of quantum theory. Before adding a few more pages to the inflatory literature on the subject, I wish to voice a somewhat sobering remark. Namely, there is a striking contrast between the amount of theoretical discussions about the nature and significance of this notion, and its practical import. Indeed, the so-called "axiom" of reduction of the state vector has a rather limited use in the effective working of quantum theory. There are very few situations where we really need to assert that, after a measurement, the state vector has been reduced to the eigenvector of the physical property corresponding to the eigenvalue resulting from the measurement ("observed"). Most of the time, in effect, the system under investigation is forgotten or discarded after the measurement - or even destroyed by it (photons). We rarely make successive observations of the same system, so that we do not usually need to know what the state of a system is after a measurement. One might argue that the reduction of the state vector is operationally used mainly to describe the preparation of a system before acting upon it, that is, to ensure that we are indeed dealing with spin up electrons or velocity selected protons such as we wish to experiment about. Anyway, the statement of reduction is to be considered more as a kind of consistency condition to be required at the beginning or end of a theoretical analysis than as an effective tool of this analysis - such as, for instance, the statement of unitary evolution, which leads to the concrete construction of the evolution operator, hence the S-matrix, cross-sections, etc. In other words, the practical status of the state vector reduction, even if it is considered as an axiom of quantum theory, is different enough from that of the constructive axioms of the theory.

Now, it is to be emphasized that, of course, everything happens *as if* the state vector indeed was reduced upon measurement. The rule is a working one ; the problem comes from its seeming inconsistent, or at least derogatory, with respect to the (other) fundamental axioms of quantum theory - in particular, the unitarity of the evolution operator. But this is by no means a unique and new situation in physical theory. For instance, and I borrow this very telling example from Everett, everything happens *as if* the Sun moved around the Earth ; this point of view moreover is, and rightly so, the one of most people, including physicists and astronomers, in ordinary life. An even more elementary example is the fact that everything happens, in our common experience, *as if* the principle of inertia were wrong : a body, put into motion, slows down and stops... Both these observational evidences - the Sun revolving around the Earth, the natural tendency towards rest of moving bodies - are not shown to be wrong by the underlying theories. On the contrary, they are reinterpreted and understood as appearances, *phenomena*, quite compatible with *a priori* conflicting views. In the same way, the question here is not to deny the reduction of the state vector, but rather to explain it away by "reducing" it from a fundamental to an auxiliary status : no longer an axiom, but a recipe - a most useful one.

I will now try to state *where* really is the problem with the reduction of the state vector, and where it is not. I would like first to dismiss two opposite ways of stating the problem, resulting from a (too narrow) focussing at one or the other end of the conventional chain system/apparatus/observer.

2. The non-problem of the system.

Looking back to the literature on the measurement problem, one is struck by the fact that it has long been (for some decades), and still is sometimes, completely entangled with more general issues about the understanding of quantum theory. In many well-known papers, the reduction of the state vector is discussed together with the so-called "wave-particle duality", "complementarity" or "uncertainty relations". That is, it is presented as a problem bearing on the very nature of quantum systems or entities. This is most clear in hidden-variable type theories, which purport to solve all of these difficulties at once by resurrecting a deterministic view of the world. For, if the random, probabilistic character of quantum theory is a provisional and superficial feature, and if there exists a deeper level at which one may recover a neo-classical description, all conceptual problems are simultaneously overcome.

However, I do believe that one should clearly distinguish between, on the one hand, questions dealing with the understanding and description of quantum entities *per se*, and, on the other hand, the problem of measurement. As is well known and rather trivial, the reduction of the state vector has no relevance for isolated systems, as long as we are concerned with their formal theoretical description . Confusion enters, however, if one takes too seriously some statements, going back to the heroic times of quantum theory, to the effect that only classical terms can be meaningfully used to describe reality. Then, of course, one is stuck with the requirement of describing the state of an electron, say, in terms of its position and its velocity. It was understan-dable in the 20's to stress the use of classical notions and to use them as far as possible, since no consistent theoretical formalism, nor physical intuition of quantum behaviour were available. However, 60 years later, the situation has radically changed, and most physicists today, ever when they still pay lip service to some Copenhagen (or so-called) orthodoxy, do speak and think of quantum systems in quantum terms. There *is* a collective craft feeling for quantum phenomena ; state vectors, superposition effects, operational magnitudes are used, formally and informally, without any respect for the bohrian *caveat* to always go back to classical concepts. At a foundational level, moreover, we now know that the correspondence principle is both dubious and useless : one does not need classical theory to build up quantum theory.

To conclude that point, the recasting of the conceptual structure of quantum theory, and the amending of its terminology (dispensing with duality, uncertainties, observables, speaking of quantons rather than (n)either particles (n)or waves, etc.) do give us, at least in my opinion, a self-contained, idiosyncratic and consistent picture of quantum systems as such, that is, isolated. In that picture, the reduction of the state vector does not enter. It is *not*, in other words, an epistemological problem.

3. The non-problem of the observer.

The standard analysis of the formal description of a measuring process, relies upon the correlation of the eigenstates of the system for the magnitude to be measured, with certain states of the measuring apparatus. If one does not appeal to the reduction of the state vector, the *global* state vector of the system and apparatus is a (vector) sum of such correlated states, in which all the eigenstates originally present (superposed) in the initial system state are still present, so that no single eigenvalue is privileged. If one sets up a new apparatus to measure the first one (or record its results), the correlation process builds on, and so on, *ad libitum*. It is known, since Von Neumann, that the reduction postulate may be invoked at any stage along the chain of successive measuring apparatus, with the same results.

This remark has led many thinkers to privilege a definite link in this chain, namely the last one, that is, the human observer, and to consider that the reduction of the state vector might be a psychophysiological process, in the brain, or, why not, a spiritual one, in the mind. After all, the argument is a simple and seemingly compelling one. It may be accepted, without too much trouble, that the combined state vector of the system and as many successive measuring apparatus as one desires, keeps building an unreduced superposition of correlated tensor product of state vectors, that is, includes all of the possible results. But when I do look at the screen of a Stern-Gerlach experiment, I definitely see one spot only, and I do not "feel" my consciousness to be split between different alternatives. It certainly looks *as if* reduction had taken place. So, in the name of the often mentional "psychophysiological parallelism", one is led to assume reduction at least (at last...) on the observer level.

However, this standpoint, in my opinion, results from a rather naive and somewhat archaic conception. Indeed, the very idea of "psychophysiological parallelism" reflects a straightforward and simplistic view of the mind ; appealing to it for unraveling the problem of quantum theory, shows a remarkable faith in a reductionist program able to bridge the gaps between the quantum physical level and the human psychological levels. Both these naive realism and reductionism have been largely undermined during the past fifty years, and I, for one, am most skeptical about the direct relevance of psychophysiological considerations for our problem.

If one does want to account for the properties of the human mind with respect to quantum observations, the least that can be asked is first to achieve the same goal with respect to classical observations. After all, nobody has ever constructed a complete "classical theory of measurement"... And for all we now know, the brain is of such a complication that, between the external stimuli it receives and our subjective perception, there is a very large distance, and even a wide gap, as far as understanding the connection is concerned. One of the best examples in case is that of *colour*, that simple - or is it ? - visual experience. It is commonly said that the colour sensation is identical with, or parallel to, the spectral distribution of light. While there is of course a tight link, many problems are to be solved before we understand this link. For instance, how does the eye-brain system manages to "see" a qualitative discretization of colour bands in a quantitatively continuous spectrum (and does not have the feeling of "pitch", as the ear) ? How are we to explain the "constancy of colour", i.e. the stable green of grass, independent of illumination (see the famous Land's experiments)? A much more elementary example is that of *speed*, which, while being a basic concept of classical mechanics (and its notion of state), by no means is an easy magnitude to observe, or measure directly... As a matter of fact, it might be said that most, if not all measurements, in effect are position measurements - were they that of a needle on a scale. And a comprehensive classical theory of measurement should give an account, both formal and constructive, of the possibility of "reducing" any measurement to positional ones.

Perhaps this may be done. But it has not been. And the question has not even been asked. Is it reasonable then, to ask from quantum theory a more direct and deeper account of the connection between its basic theoretical concepts and the final observational effects, than has ever been established for classical theory ?

I would conclude, then, that we have better leave our brains outside our area of interest as physicists, and that it is preposterous to hope for an explanation of reduction at that level. In other words, it is *not* a psychophysiological problem.

4. The problem of apparatus.

If the problem is not an epistemological nor a psychophysiological one, perhaps it is a *physical* one ? If it is not the system, nor the observer which are to be questioned, it remains but the apparatus to understand.

And, indeed, it is known, but not yet fully appreciated, that for an apparatus to play its role, that its own eigenstates must be orthogonal, so that, upon measurement, any interference terms between eigenstates of the system are washed away. The full state-vector of the combined supersystem (system + apparatus) then is trivially equivalent to a density matrix for the system, that is, everything happens *as if* the state vector of the system had been reduced.

But that is the beginning, and not the end of the story... What remains to be accomplished, and that is the purpose of the various contributions to this book, is to show that such apparatus may be constructed in accordance with the very laws of quantum theory. This question may be analyzed under different angles, from logical/terminological ones : what kind of system are we entitled to call a measuring apparatus, and what questions are we allowed to ask about it, to empirical ones : how to characterize and build a macroscopic quantum system ? In fact, these are subquestions of a larger one which encompasses the so-called theory of measurement, and that is : how comes that a large quantum system sometimes (in fact, often) simulates so well a classical one ? In other words, how are we to derive and justify the validity of the classical approximation to quantum theory ? It is remarkable enough that we know so little about the emergence and validity of classical theories from the underlying, and supposedly deeper, quantum theory. Beyond superficial considerations ($h \rightarrow 0$, Ehrenfest's theorem), the problem is *not* a simple one, as proved by the existence of macroscopic quantum systems, and by the duality of classical physics (wave *or* particle theories).

The final understanding of the reduction recipe probably requires a clarification of the foggy borderline between the quantum and classical realms, as the measurement process precisely is concerned with the crossing of this border. This clarification, it must be stressed, requires, along with a theoretical analysis, a careful terminological assessment. For instance, much harm has been done against the no-reduction position advocated here, by its hasty labeling as the "many universes" interpretation. Indeed, the latter amounts to mix up a quantum superposition of quantum correlated states with a coexistence of classical states ; to take a quantum *plus* for a classical *and* precisely begs the question under discussion.

5. Conclusion : the classical problem.

To the initial question "where is the problem (with the reduction of the quantum state vector)?", I would therefore venture the paradoxical answer : "in *classical* theory"... This is true, I hold, at the specific level of the understanding of measurements as physical process, and apparatus as physical systems - that is, quantum ones, but classically simulated. It is also true at the general level of the epistemological status of quantum ideas. For let us go back to the analysis I have proposed of the two non-problems, that of the system and that of the observer.

Both the question of the conceptual nature of physical entities and that of the relationship between basic notions and direct perceptions, have haunted the development of quantum theory. But both these questions might be asked, and should be, already for classical physics. Even if they are technically easier to analyze, their philosophical depth is the same. However, throughout most of its history, and in particular during the 19th century, classical physics has been protected, so to speak, from these inquisitive discussions, by the rather naive faith of most physicists, if certainly not of philosophers, in a realist and reductionist epistemology. In some sense then, quantum theory has now to pay the debt of its prodigal ancestor. That may account for our reluctance...

References

The literature on the subject is so vast that, in a nontechnical paper such as the present one, any specific bibliography is doomed to be incomplete, or even meaningless. The interested reader no doubt already knows the main papers in the field and general sources such as the extensive anthology by J.A. Wheeler and W.H. Zurek eds., *Quantum Theory and Measurement* (Princeton University Press, 1983), or the careful historical study by M. Jammer, *The Philosophy of Quantum Mechanics* (Wiley, 1974).

For the same reason, I have chosen not to make specific references to my own papers on the subject which, however, give more detailed arguments on most of the points too briefly treated here. See, for instance, my articles :

"Les inégalités de Heisenberg", *Bull. Soc. Fr. Phys., Enc. Péd.* **1** (1973)

"The pedagogical role and epistemological significance of group theory in quantum mechanics", *Riv. Nuovo Cimento* **4**, 99 (1974)

"Quantum fact and classical fiction", *Am. J. Phys.* **44,** 130 (1976)

"Towards a proper quantum theory", in *Quantum Mechanics, Half a century later*, J. Leite Lopes & M. Paty eds. (Reidel, Dordrecht, 1977), p.171

"The picture of the quantum world: from Duality to Unity", *Int. J. Quant. Chem.* **12,** supp.1, 415 (1977)

"Classical apples and quantum potatoes", *Eur. J. Phys.* **2**, 44 (1981)

"Quantum physics and language", *Physica B* **151**, 314 (1988)

"A quantum credo", *Physica B* **151**, 378 (1988)

"Neither waves, nor particles, but quantons", *Nature* **334,** 6177 (1988)

Let me mention also the textbook in quantum physics written by myself and Françoise Balibar :

Quantique (Rudiments) (Interéditions/CNRS, 1984). English translation, *Quantics (Rudiments)* (to be published, North Holland, 1990)

Experiment and Quantum Measurement Theory

Anton Zeilinger

1. INTRODUCTION

An experimentalist trying to work his way through the impressive body of literature in quantum measurement theory rapidly notices that the theory, at least hitherto, does not make any predictions for experiments aimed at its own test. Rather one has the impression that most experiments discussed in the context of that theory serve more the purpose of trying to demonstrate its internal consistency. An early example is the beautiful paper by Mott (1929) where he presents the wave mechanical analysis of a seemingly particle-like phenomenon, the α-ray tracks in a cloud chamber.

This situation has also been fostered by the epistemological complexity of quantum mechanics and the question of internal consistency of the theory has been the focus of numerous gedanken experiments since its early days, the most notable ones being those proposed by Einstein in his famous dialogue with Niels Bohr (Bohr 1949).

Up to this day the epistemological problems of quantum theory have certainly not been mastered yet as evidenced for example by the following quotations of two scientists who had made significant contributions to the theory. Feynman (1967) writes "I think I can safely say that nobody understands quantum mechanics" and Penrose (1986) expresses similar concern when he states that standard non-relativistic quantum mechanics: "... makes absolutely no sense."

Certainly a most important point is that quantum theory is supported by a vast body of experimental data constituting an extremely impressive agreement with theoretical prediction. In view of that fact it might at first impression seem superficial to request further experiments to be performed. Yet we note the interesting phenomenon that only very little experimental activity is aimed at testing quantum measurement theory itself. Even worse, no experimentalist in his laboratory needs to use the full quantum measurement theory in order to interpret his experimental results. Therefore, extremely speaking, one could have the impression that quantum measurement theory might be the only physical theory no experimentalist needs.

The main point of the present chapter is that recent technological progress makes it possible to do experiments in the laboratory which, until now, have been confined to the realm of gedanken experiments. It therefore is timely to ask which kind of experiments on quantum measurement theory would be most useful and what aspect of the theory might need further testing.

When the experimentalist studies quantum measurement theory he immediately notices significant differences between the way he would describe measurement and the way the theoretical physicist does. This observation may be related to deficiencies in the communication between the two groups of physicists. The communication seems to work more or less satisfactorily on the level of an exchange of the results between the two groups but it is much worse with respect to an in-depth understanding of the methods used by either side. It is even worse when it comes to more ephemerous manifestations of scientific activity like an individual scientist's motivation, his goals or even his philosophical attitude.

At first impression these points might seem superficial but, I propose, they are intricately interwoven with any scientist's activity and they may even significantly influence his directions of work. I consider quantum measurement theory to be a case at point and I argue that the philosophical attitude of most experimental physicists results from a very operational approach to experiment. That approach, though certainly rather successful in the laboratory, might limit the scope when it comes to identifying new directions of research. Along the way I shall give a down-to-the-Earth analysis of the measurement problem and eventually I shall propose that one of the most interesting challenges for today's experimentalist lies in demonstrating quantum behaviour for ever larger objects.

2. THE EXPERIMENTALIST'S OPERATIONAL PHILOSOPHICAL ATTITUDE

Anybody visiting a modern physics laboratory certainly notices immediately the high level of technological sophistication of the equipment used and he may readily conclude that such work has to be complemented by significant financial expenditure. In fact, modern physics research laboratories very often are quite similar in appearance to modern high-tech companies and therefore we might very well conjecture that the purpose of such laboratories is to be factories of knowledge.

This impression is supported by the responses given by experimentalists to questions concerning the reason for their endeavour. One purpose often given is to advance the knowledge of mankind. This is then specified as the aim to test a certain theory or, more modestly, to verify some theoretical predic-tion. Also, the development of new technology is often mentioned as being the aim of a physicist's work. Specifically, any hint at possible development of new materials generally is accepted to provide enough justification for a certain line of work. Though these motivations regularly are also those presented to funding authorities and agencies, I may submit here that they rarely are the true motivations of experimentalists and in even rarer cases do they constitute their primary motivation.

Notwithstanding the importance such worldly goals like reputation, career, power or income might have in the motivation of the individual scientist, I would like to point out that there is another, possibly more important, stimulus rooted in the personality of the experimentalist. Hints can be obtained in conversations with experimentalists where frequently formulations like "try out something new" or "play a new game" are used. There is also strong emphasis on the "fun" provided by mastering experimental challenges nobody else has mastered before.

It is therefore my impression that satisfaction of the playful traits of *homo faber* regularly provides the main motivation for experimentalists to invest not only their energy but often the better part of their lives into meeting experimental challenges. As a corollary, experimental ideas are, at least in their early stages, primarily judged on the basis of the challenge they provide and of their feasibility. This attitude, being quite successful in the daily routine work in the laboratory, is also the major motor for technological progress.

In contrast, the question as to what the final goal of a certain line of research and particularly the possible applications of the results might be regularly arises primarily because of the necessities to justify to the society the significant expenditures. I feel very strongly that this necessity for justification creates a rather deplorable state of affairs. Not only is it deplorable because it may reasonably be conjectured that it provides an incentive for presenting exaggerated claims, it is also a rather counterpro-ductive enterprise. There are numerous examples in the history of physics where not only nobody at the time had any reasonable feeling of the possible technological applications of a new experimental success but where it was completely impossible to make any reasonable technological forecast.

This again is connected to observations like the fact that for many applications the technological environment in a society has to be at a certain critical level or that some applications often require as much inventiveness and ingenuity as the original idea did. An explicit example is provided by the experiments performed by Heinrich Hertz in the Eighteen-Eighties. These experiments were considered to be fundamentally important because they would provide the final confirmation of Maxwell's equations but at that time it was impossible to even moderately guess the wealth of applications springing from Hertz's work. Another famous example is the laser, which, for quite some time after its invention, was considered by many to be the "perfect solution looking for its problem".

I would like to conclude this paragraph firstly with the remark that the question "What is it good for" does not interest most scientists working in basic research and rightly so. It may reasonably be conjectured that this question has already caused · enormous damage by discouraging ideas to be pursued and therefore, at least in the field of basic research, it should never be asked.

Secondly, I wish to propose that the operationalistic attitude identified above in the experimentalist's motivation has its counterpart in his approach to epistemological and interpretive questions. There, to my impression, the primary reason for accepting or discarding a certain position again is "whether it works or not" in the day-to-day routine environment of a laboratory. Issues of logical consistency appear to be much less important. We will show in the next paragraph what this implies for the quantum measurement problem and in which way that attitude might be the cause of limitations in a scientist's scope when contemplating new experiments.

3. AN OPERATIONAL APPROACH TO QUANTUM MEASUREMENT THEORY

Having identified in the last paragraph the rather operational attitude of the experimentalist, I set out in the present chapter to view quantum measurement theory from that perspective. In analogy, my observation here is that the experimentalist's approach is signified by the prime criterion of whether something "works" and that questions of internal consistency or of the philosophical consequences of a certain point of view are regarded to be secondary if not unimportant at all.

It is well known that the main problem of quantum measurement theory results from the linearity of the theory together with the superposition principle. This point has been analysed by many since the early days of quantum physics as witnessed for example by the compilation arranged by Wheeler and Zurek (1983) of most of the relevant papers in the field. Following von Neumann (1932) a most transparent presentation has been given by Wigner (1963) and we will discuss here a very much abbreviated version.

It is certainly the primary purpose of a measurement to obtain some information on some properties of a system which may be represented by a state $|\psi>$. Suppose we do have at our disposal a measurement apparatus whose state is designated as $|\Phi>$. Then, in order to obtain information on $|\psi>$ through observations of the apparatus $|\Phi>$ we have to establish some correlation between the system and the apparatus. For simplicity let us contemplate nondegenerate discrete variables only. This is usually meant to imply that there is a pointer on the apparatus with a finite set of discrete and well-distinguished positions. The states of the apparatus with well-defined pointer positions may be designated as $|\Phi_i>$ (i = 0, 1, 2, ...) where i = 0 refers to the situation at the beginning of the experiment where no pointer reading has been obtained yet.

Let us first analyse the simple situation when the system is a state that gives a unique well-defined reading of the pointer (usually called an eigenstate $|\psi_j>$ of the observable defined by the apparatus). Then the establishment of the correlation part of the measurement may be described as the transition

$$|\psi_j> \ |\Phi_0> \ \Rightarrow \ |\psi_j> \ |\Phi_j> \tag{1}$$

Observation of the apparatus now leads to the well-defined answer that the pointer is found to be in position j. Though this statement usually is regarded to be rather safe from criticism we might note here that its operational establishment is far from trivial. It is evident that it can only be established through observation of a large ensemble of identically prepared systems and it is this feature which, to my opinion, already leads to interesting questions not often analysed in the literature.

The first problem refers to what we mean by "identically prepared". One might easily be tempted to assume that the way to establish whether N systems of an ensemble are identical would be to subject them all to the same measurement procedure obtaining always the same result. Yet such a position, I submit, would directly lead to a logical circle in defining measurement on the basis of the concept of identically prepared systems, the latter concept again being defined only through the results of measurements, the concept to be defined in the first place.

Though from certain viewpoints this might be acceptable, I would rather propose here to base the concept of identical preparation on a more worldly footing. Evidently, for an experimentalist, systems are prepared identically if, to the best of his knowledge, they emerge from the same apparatus used for their preparation with none of the properties of the apparatus changed. Without entering now into a discussion of the, by far not trivial, meaning of this last clause, it is evident that therefore we never can be certain whether or not all systems are really prepared such as to be in an eigenstate of the observable defined by the measurement apparatus.

Another important feature of the analysis so far is the property that we did base our concept of identical preparation on the much more fundamental concept of information ("... to the best of his knowledge ..."). Any firm foundation of quantum theory eventually has to make recourse to information theory in some way. The most important contributions so far in that direction are those by Wheeler (1986), by Wooters (1980), by von Weizsäcker (1985) and by Summhammer (1988).

In order to proceed with our analysis of quantum measurement theory we assume now that the system initially is in a superposition

$$|\psi\rangle = \sum_i g_i |\psi_i\rangle \qquad (2)$$

of different eigenstates $|\psi_i\rangle$. Here, g_i represent the respective complex amplitudes.

This latter sentence again has a rather complicated operational meaning. For the experimentalist it is connected to a large set of instructions as to how to establish superpositions of this kind. Particularly, as has been illustrated beautifully by Feynman *et al* (1963) for the case of two- and three-component spin variables, the necessity to give meaning to the complex amplitudes g_i implies a set of different apparatus settings with the trans-formation between them being represented by unitary operations.

Though this point certainly deserves more attention, we will subject our system (2) now to the measuring apparatus thus obtaining the superposition

$$\sum_i (g_i |\psi_i\rangle) |\Phi_0\rangle \Rightarrow \sum_i g_i (|\psi_i\rangle |\Phi_i\rangle) \qquad (3)$$

which is a direct consequence of the linear superposition principle. For many theoretical colleagues the measurement is completed as soon as the statistical correlation described by (3) has been established (see *e.g.* Wigner 1963). For the experimentalist such a position is again unacceptable because so far we have not obtained any information yet.

In order to obtain this information we have to read the pointer position and therefore we have give operational meaning to the right hand side of equation (3). To do this we use of the projection postulate of quantum mechanics. Accordingly, we will always find the pointer in a well-defined position j. We may then conclude that the system and apparatus together are in the product state $|\psi_j\rangle\ |\Phi_j\rangle$ and therefore the system itself is in a well-defined state. Yet it is evident that for the individual system that state is only well-defined after the measurement and not before when the state is still the superposition (2). For a particular experiment the meaning of the weights g_i is only that the absolute square $|g_i|^2$, if properly normalized, gives the probability for an individual outcome i. The phases of the weights g_i only have meaning in more complicated experiments when we, in some way or other, change the setting of our apparatus.

Importantly, a problem then arises as to whether the superposition (3) is more than just a representation of the individual outcomes i together with their relative weights $|g_i|^2$. In particular, as has first been signified by the famous cat paradox (Schrödinger 1935), equation (3) would imply that the measurement apparatus is in some superposition of different pointer states if the system is in a superposition to begin with. Now, whether or not this presents a problem certainly depends on the position one assumes as to what constitutes a measurement apparatus. If, as done in quite a number of theoretical analyses of the problem, one is readily willing to give apparatus status to microscopic variables like the position of a particle in a Stern-Gerlach magnet no problem arises yet, at least at that level.

Yet, as I might note here, no experimentalist of my personal acquaintance, when prompted, is ready to admit that such microscopic variables justly might be named measuring apparatus. Rather, to the experimentalist, a measuring apparatus always has to include all the hardware necessary to actually read out the information in some way. It then does not make any sense to the experimentalist to talk about quantum states of such an apparatus. He simply has never seen a measurement apparatus in his laboratory for which he had to assume the existence of superpositions of pointer states. For example his computer printer putting down the results of his experiments permanently onto a piece of paper works perfectly well as a classical machine.

Therefore to the experimentalist this part of the quantum measurement problem simply does not exist. In fact, it is my own personal experience that it is very difficult, if not impossible, to convince an experimentalist that there might be a problem at all lurking behind the question of what super-positions of pointer states might mean. If anything, he would readily assume that one has to run into a problem if one tries to describe a macroscopic apparatus with quantum laws, a procedure which, to him, is clearly unnecessary and nonsensical.

4. IS QUANTUM THEORY UNIVERSALLY VALID?

The experimentalist, rejecting the notion that a macroscopic measurement apparatus should be described quantum mechanically, usually does not realize that in assuming such a position he implicitly negates the universal validity of quantum mechanics. This is certainly unacceptable from a purist's point of view because on the one hand it would mean that we might have to abandon present quantum mechanics as a universal theory and on the other hand all experimental evidence so far suggests that there are no limits to the validity of the theory.

One might nevertheless very well question the validity of the claim that quantum mechanics is universally valid. Upon close inspection it is rather obvious that such a statement is not an unescapable consequence of experimental observation at least to this date. Though no experiment has yet found any contradiction to quantum mechanics, no experiment so far has demonstrated that in order to describe, for example, the motion of an automobile we have to invoke any quantum mechanical laws. It is very often proposed that such would not be sensible anyway because, even if we were successful to prepare a macroscopic object in a superposition of the type (2), we would find out that in the density matrix description any off-diagonal elements $g_i g_j$ would rapidly fade away for a variety of reasons.

As has been analysed in detail by Leggett (1987) this solution of the measurement problem is not acceptable because it eschews the question whether the macroscopic apparatus is in a superposition or not. Even worse, it assumes implicitly conceptual equivalence between a classical ensemble description and a quantum density matrix description. In the classical description each member of the ensemble has definite properties while in the density matrix case the members of the ensemble are not in definite states as long as the off-diagonal matrix elements are nonzero, no matter how small they might be in a particular situation. While it is true that a classical ensemble may be described using diagonal density matrices, the problem clearly arises at the point of transition from the full density matrix description of a collection of individual quantum systems to a description with a diagonal density matrix only. There, to me, a giant interpretive leap happens as signified for example by the problem that then any individual system has somehow to make the transition from being in a superposition of the density matrix base states into only one of these states.

To the contrary, it might very well be possible that we have to give up the goal, so successful in the history of physics so far, to reduce all laws on a more complex level to a few fundamental ones. In particular, in the case of quantum measurement theory, it probably will never be possible to arrive at a complete quantum mechanical description of a macroscopic object as large as our automobile. This follows from the observation that in order to describe all internal degrees of freedom of such an object consisting of the order of 10^{25} constituent particles we would need more information storage capacity than we will ever have at our disposal even if we strained our phantasy to its utmost extreme. Since a full quantum description of macroscopic objects including all internal variables does not exist and will never exist the claim that such a description is thinkable in principle is devoid of operational meaning. An experimentalist will always request to be given a set of instructions telling him how to verify in the laboratory a specific theoretical concept.

I would like therefore to conclude the present paragraph by proposing that it is an explicit challenge to perform experiments testing the proposition of the universal validity of quantum mechanics and that the experimentalist should not let his prejudices as to the non-quantum nature of macroscopic objects be in his way in planning such experiments. While I argued above that we will never achieve a detailed quantum mechanical description of macroscopic objects, we might very well achieve experimental confirmation of many phenomena of collective quantum behaviour.

5. THE COLLAPSE OF THE WAVE PACKET

Directly connected to the measurement problem is the question of the nature of the collapse of the wave packet. Put in simple terms, the question arises whenever we do a measurement on a system described by a superposition as given in equation (2). We then find the system in one of the eigenstates of the operator defined by the measuring apparatus. A specific example is the case when we are actually considering a wave packet as being composed as a superposition of plane waves. Such a wave packet is more or less well-localized but we always can perform position measurements on that wave packet which are better localized than the dimension of the packet itself.

As an example consider our recent series of experiments with beams of very cold neutrons (Eder *et al* 1989, Gruber *et al* 1989). In these experiments we regularly work with a neutron beam of average deBroglie wavelength of 102Å corresponding to an average velocity of 39 m/s. The wavelength bandwidth of the beam is 22 %. The experiments are done on an optical table permitting neutron flight paths of up to 6 meters. It is easy to see that if we start initially with a minimum wave packet for the incident neutrons, such a wave packet over the full 6m path length spreads out to a length of the order of 1 meter. Our neutron detectors typically have a diameter of about 2 cm. Hence, registration of a neutron in the detector results in a localization which is significantly better defined than the dimension of the wave packet.

This phenomenon has first been put forward as a strange feature of quantum mechanics by Einstein in his famous dialogue with Bohr (1949) and it might certainly appear to be miraculous if we, as many physicists do, should picture the deBroglie wave as a real wave spread out in space. Then such a wave would have to collapse instantly all over space. Yet, as we will argue in the next paragraph, such a realistic picture of the deBroglie wave should not be taken seriously anyway no matter how suggestive diffraction patterns like the one shown in Fig. 1 might be.

Part of the problem disappears if we consider the wave function to be just a representation of our knowledge of the system. Then it is not surprising that the wavefunction changes if we do a measurement, i.e. if we gain some new information. Other aspects of quantum measurement still remain to be problematic then because the wavefunction represents the maximum possible knowledge of a system. Therefore, unlike in classical mechanics, a measurement in quantum mechanics does not just add some new knowledge, it changes our knowledge in a strange way. We may gain some new information about some parameters at the expense of irrecoverably loosing information on some others. This point has first been analysed by Schrödinger (1935) in his series of papers on the situation of quantum mechanics. In the meantime we have learned, particularly through Bell's theorem, that we should not even think of properties of a system being definable that are not specified by the wave function. In that sense the reduction of the wave packet probably will never submit to a successful explanatory reduction to basic principles.

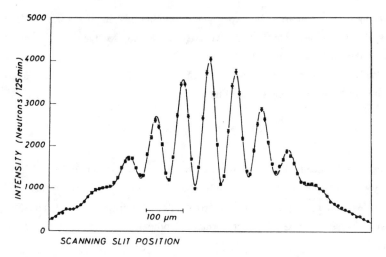

Fig. 1: Distribution of very slow neutrons after diffraction at a double-slit assembly. The solid line represents the prediction of standard quantum mechanics (for details see Zeilinger *et al* 1988).

6. WHAT IS THE deBROGLIE WAVELENGTH?

Though the deBroglie wavelength is a theoretically well-defined entity, one faces considerable problems when trying to understand it conceptually. The problems partially result from the fact that the deBroglie wavelength seems to be a concept which can easily be used in obtaining a picture-type imagination of an experiment. Yet, things are not that simple. Besides the problems related to the reduction of the wave packet there are some interesting features of the deBroglie wavelength which, to my opinion, not just limit its usefulness for obtaining a pictorial representation, they make such a representation false and therefore dangerously misleading.

The first problem arises when we realize that the deBroglie wavelength is not gauge invariant. In fact, all predictions remain unchanged if we simply substitute

$$\mathbf{k'} = \mathbf{k} + \mathbf{q} \qquad \text{with} \qquad \oint \mathbf{q} \cdot d\mathbf{r} = 0. \tag{4}$$

This is related to the property that the deBroglie wavelength is only defined through interference experiments, i.e. when at least two paths, albeit possibly neighbouring ones, interfere. Thus, strictly speaking, the deBroglie wavelength only appears in experimental predictions in the form of a path integral over closed loops $\oint \mathbf{k} \cdot d\mathbf{r}$. At least to my knowledge there is no other way to define it operationally.

Another problem is that the deBroglie wavelength is not Galilei invariant. To the contrary, it changes according to

$$\mathbf{k'} = \mathbf{k} + m\mathbf{v}/\hbar. \qquad (5)$$

It has already been nicely pointed out by Levy-Leblond (1974 and 1976) that the wave-function ψ and the deBroglie wavelength λ have rather strange properties under simple Galilei transformations, properties not found in any classical wave. The Galilei transformation property (5) is a corollary of (4) because

$$\oint m\mathbf{v} \cdot d\mathbf{r} = 0. \qquad (6)$$

None of these properties are shared by classical waves and therefore the deBroglie wave does have no conceptual significance. It merely evidences itself as an aid to calculating interference patterns, which again means that it only helps us to calculate statistical predictions of the distributions of particles on an interference screen.

As an example let us briefly consider again the measured distribution of neutrons (Fig. 1) after diffraction at a double-slit assembly (Zeilinger *et al* 1988). We might notice some interesting features of that diffraction pattern. If we briefly draw our attention to the scale of the intensity we notice that this is at most of the order of a single neutron per second. Therefore, since the neutrons used in that experiment had a velocity of 200 m/s we may readily conclude that, while one neutron is being registered, the next one to be counted in most cases is still sitting in its U-235 nucleus waiting to be born as a free neutron. Secondly we may note that the diffraction pattern changes completely if we close one of the two slits (Fig. 2). The important point is now that there are positions on the screen where closing one slit increases the number of neutrons arriving there. This is clearly an explicit demonstra-tion of destructive interference and, to me, it demonstrates that the wave function somehow has to be associated with each individual neutron and not just with the ensemble. We have to accept that interpretation in order to get some understanding of the fact that each neutron has a reduced tendency to land at certain positions if both slits are open.

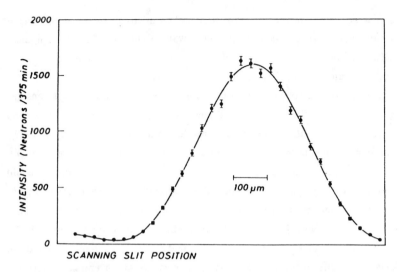

Fig. 2: Diffraction of very slow neutrons at a single slit.

7. ENTANGLEMENT AND COMPLEMENTARITY

It is generally accepted that the double-slit interference pattern, and for that matter any interference pattern, only arises if we have no way of telling which path the particle took. Discussions of this point are as old as quantum theory itself (Bohr 1949). Yet there are still a number of misconceptions being put forward again and again. The most significant misconception arises again from a certain operational point of view. According to that view, the complementarity between interference pattern and information about a particle's path arises from the fact that any attempt to observe the particle path would be associated with an uncontrollable transfer of energy and/or momentum to the particle. Such an energy/momentum transfer in itself would then be the reason for the loss of the interference pattern. This viewpoint goes back at least as far as the original discussion of the so-called γ-microscope by Heisenberg (1927) and it may for example be found in the discussion of the double-slit experiment by Feynman (1963). In this view quantum uncertainty arises because of unavoidable disturbances which again are due to the intrinsic clumsiness of any macroscopic measuring apparatus.

In the present paragraph I shall argue that the situation is not that simple and I shall discuss possible relevant experiments. Experimental progress over the last few years has made it possible to consider actually feasible experiments on complementarity quantum interference.

One line of such research considers the use of micromasers in atomic beam experiments (Scully *et al* 1989, Scully and Walther 1989). Typically, in such an experiment, an atom passes through a cavity such that it exchanges exactly one photon with the cavity. Thus, by investigating the cavity, one has information on whether or not an atom passed through it. Now, if we consider placing one cavity each into the two beams of an atomic-beam interferometer, we may obtain information on which path the atom took (Scully *et al* 1990). This can in principle be done at a time after the atom has been registered on the interference screen and hence, we conclude, no interference pattern should arise. On the other hand, we can read the information in the micromasers in such a way as to obtain no information on which micromaser the photon had been stored in. Then we have just the information that the atom passed through the apparatus but not along which path. In that case the atoms counted in coincidence with the photons should again be members of an ensemble defining an interference pattern. Such experiments, being already conceptually quite remarkable are certainly within reach of current technology and within a few years we should witness their experimental realization.

Another independent approach to complementarity in an interference experiment is opened up by the correlated photon states emerging from nonlinear crystals through the process of parametric down-conversion (Burnham and Weinberg 1970). The principle of a typical experimental setup is shown in Fig. 3. There, the beams originating from a parametric-down conversion crystal operating in the non-degenerate mode feed two Mach–Zehnder interferometers (Horne *et al* 1989). For the purpose of our present discussion we need not to go into the details of the parametric down-conversion process. It suffices to know that if certain experimental conditions are met the photon pairs emerging from the crystal source S are found in the state

$$|\psi\rangle = 2^{-1/2} \left[|A_1\rangle|C_2\rangle + |D_1\rangle|B_2\rangle \right] \tag{7}$$

where, *e.g.*, the ket $|A_1\rangle$ describes the fact that photon 1 is found in beam A. States of the type (7) have been called by Schrödinger (1935) "Entangled States" (in German "Verschränkte Zustände") because they describe a close interconnection between the properties of the two particles. In detail, using the reduction postulate, state (7) predicts that whenever we find photon 1 of the pair in beam A, we find its counterpart in beam C and whenever we find photon 1 in beam D, photon 2 is in beam B.

We may then conclude that no interference pattern should be observable upon variation of either the phase ϕ_1 or the phase ϕ_2. This can easily be

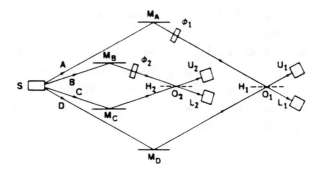

Fig. 3: An arrangement for two-particle interferometry (Horne *et al* 1989). The source S emits two photons in the entangled state (6). Photon 1 traverses the Mach-Zehnder interferometer starting with the beams A and D while photon 2 traverses the Mach-Zehnder interferometer starting with the beams B and C. Phase shifters in both interferometers permit continuous variation of the phases ϕ_i.

calculated starting from state (6) and taking properly into account the properties of the half-silvered mirrors. Yet, without going into the details of the calculation, we readily see why this has to be the case. Assume, for example, that we would obtain an interference pattern for photon 2, *i.e.* a sinusoidal variation of the intensities registered in the detectors U_2 and L_2 upon variation of the phase ϕ_2. Then we could independently determine whether photon 1 of any given pair is in beam A or in beam C. This, by virtue of the entanglement expressed in state (6), would permit us to conclude for each and every individual photon 2 whether it took path B or path C. Thus we would have both the interference pattern and the path information at the same time. To avoid that problem our conclusion, supported by calculation, is that there never arose an interference pattern in the detectors U_2 and L_2 to begin with.

Yet, if we recombine also the two paths of particle 1 as indicated in Fig. 3, and if we register both particle 1 in either detector U_1 or detector L_1 and particle 2 in either detector U_2 or L_2 we have forgone any possibility of obtaining path information. Therefore, we conclude, an interference pattern should arise in coincidence counts between the detectors for particle 1 and for particle 2 shown in Fig. 3. This indeed follows from the quantum mechanical calculation (Horne *et al* 1989). Experiments of this type are under progress both along the lines discussed here (Rarity and Tapster 1989) and along similar lines involving interferometers with single input beams only (Franson 1989, Kwiat *et al* 1990).

Concluding the present paragraph I would like to stress that we have discussed here experiments where the reason why no interference pattern arises is not any uncontrollable disturbance of the quantum system or the clumsiness of the apparatus. Rather it is the fact that the quantum state is prepared in such a way as to permit path information to be obtained for both particles simultaneously by measuring one particle only. We can regain the interference pattern by actively destroying any possibility of determining either particle's path.

8. REVISITING THE MEASUREMENT PROBLEM

After all the, admittedly very subjective, discussion given above I would like to return now again to the question as to which light future experiments could possibly shed on the measurement problem and on the interpretation of quantum mechanics. This question is inherently impossible to be answered in a way today which will stand to scrutiny by future generations of physicists. Or, as one might put it in simpler terms: "It is very difficult to predict, particularly the future." Nevertheless we might legitimately use quantum measurement theory itself to help us identify possible interesting avenues for experimental research. This way we might mollify the harsh statement shared by many of my experimentalist colleagues that quantum measurement theory is the only physical theory no experimentalist needs to allude to in order to understand, interpret or design his experiments.

One of the rather certain points is that in the not-too-distant future we will succeed in performing really detailed complementarity experiments. This will include experiments that will demonstrate explicitly how the interference pattern in a double-slit experiment disappears as soon as we have the possibility of obtaining path information and also how it reappears if we decide to forego any such possibility. It will be very interesting to demonstrate exactly that this is not a sharp Yes-No transition but a continuous one with the possibility of having both partial path information and a somewhat reduced interference pattern at the same time (Wooters and Zurek 1979 and Zeilinger 1986).

Another very promising avenue results from the implicit claim of orthodox quantum measurement theory that quantum theory should be universally valid and that therefore quantum theory should *in principle* describe also the behaviour of macroscopic objects. In that sense it is a challenge to the experimentalist to demonstrate quantum effects for increasingly larger objects.

There are two very fruitful avenues of this kind being already pursued today by experimentalists. One is the series of experiments on the macroscopic phase coherence of superconductor currents (see *e.g.* Clark 1987). There we typically have a huge number of Cooper pair electrons in one quantum state and that state occupies a sample of up to centimeter size.

An equally interesting and challenging direction of research is to show deBroglie wave behaviour for increasingly larger objects. Particularly convincing are interferometry experiments of this kind. We may note that in neutron interferometry we for the first time had already a wealth of experiments confirming quantum behaviour of a non-elementary, composite particles (see for example Badurek *et al* 1988). While there the number of constituents (three quarks, not counting the gluons) is still rather low, this will significantly change with interferometers for atoms or even molecules. In that case we will achieve verification of quantum laws for objects consisting of the order of 100 to 1000 constituents. After the beautiful demonstrations of atomic beam diffraction at crystals (Knauer and Stern 1929, Estermann and Stern 1930, Estermann *et al* 1931), at standing light waves (Martin *et al* 1988) and at transmission gratings (Keith *et al.* 1988) there is no doubt that such experiments shall provide a wealth of evidence for the validity of quantum mechanics at that level.

An interesting common point to all these experiments is that for their description we do not need to solve the multi-particle Schrödinger equation containing all constituents of the object (supercurrent, neutron, atom, molecule) explicitly. Rather it suffices to use bulk properties like, in the case of atomic beam diffraction, the total mass of the atom in order to calculate the deBroglie wavelength. We may therefore reasonably conjecture that at present no maximum size of an object for quantum mechanics to be valid is anywhere in sight. Yet we are certainly aware of the fact that the larger an object, the better we have to isolate it from environment in order not to have to deal with excitations of its internal degrees of freedom due to interactions with the environment. This, at least at present, seems to be more a question of effort, or financial commitment for that matter, than of principle.

Acknowledgment

The work with very cold neutrons reported here has been supported by the Austrian Fonds zur Förderung der wissenschaftlichen Forschung under project No. 6635.

References

Badurek G, Rauch H and Zeilinger A 1988 eds *Proc Int Workshop on Matter Wave Interferometry* (Vienna) *Physica B* **151** pp 1-400

Bohr N 1949 *Albert Einstein, Philosopher Scientist* ed P A Schilpp (Evanston: The Library of Living Philosophers) pp 200-41

Burnham D C and Weinberg D L 1970 *Phys Rev Lett* **25** 84

Clark T D 1987 *Quantum Implications* ed B J Hiley and F David Peat (London and New York: Routledge and Kegan Paul) pp 121-150

Eder K, Gruber M, Zeilinger A, Gähler R, Mampe W and Drexel W 1989 *Nucl Instrum Methods* **A284** 171

Estermann I and Stern O 1930 *Z Physik* **61** 95

Estermann I, Frisch R and Stern O 1931 *Z Physik* **73** 348

Feynman R P, Leighton R B and Sands M 1963 *The Feynman Lectures of Physics* (Reading Massachusetts: Addison-Wesley) vol III

Feynman R P 1967 *The Character of Physical Law* (Cambridge Massachusetts: MIT Press) p 129

Franson J D 1989 *Phys Rev Lett* **62** 2205

Gruber M, Eder K, Zeilinger A, Gähler R and Mampe W 1989 *Phys Lett A* **140** 363

Heisenberg W 1927 *Z Physik* **43** 172

Horne M A, Shimony A and Zeilinger A 1989 *Phys Rev Lett* **62** 2209

Keith D W, Schattenburg M L, Smith H I and Pritchard D E 1988 *Phys Rev Lett* **61** 1580

Knauer F and Stern O 1929 *Z Physik* **53** 766 and 779

Kwiat P G, Vareka W A, Hong C K, Nathel H and Chiao R Y 1990 *Phys Rev A* (in press)

Leggett A J 1987 *Quantum Implications* ed B J Hiley and F David Peat (London and New York: Routledge and Kegan Paul) pp 85-104

Levy-Leblond J-M 1974 *Riv Nuovo Cimento* **4** 99

Levy-Leblond J-M 1976 *Am J Phys* **44** 11

Martin P J, Oldaker B G, Miklich A H and Pritchard D E 1988 *Phys Rev Lett* **60** 515

Mott N F 1929 *Proc Roy Soc* **A126** 79

von Neumann J 1932 *Mathematische Grundlagen der Quantenmechanik* (Berlin: Springer)

Penrose R *Quantum Concepts in Space and Time* ed R Penrose and C J Isham (Oxford: Clarendon Press) pp 129-149

Rarity J G and Tapster P R 1989 manuscript to be published

Schrödinger E 1935 *Naturwissenschaften* **23** pp 807-812 823-828 844-849

Scully M O, Englert B-G and Schwinger J 1989 *Phys Rev A* **40** 1775

Scully M O and Walther H 1989 *Phys Rev A* **39** 5229

Scully M O, Walther H, Englert B-G and Schwinger J 1990 *Phys Rev A* (in press)

Summhammer J 1988 *Found Phys Lett* **1** 3

von Weizsäcker C F 1985 *Aufbau der Physik* (München: Carl Hanser)

Wheeler J A 1986 *New Techniques and Ideas in Quantum Measurement Theory* ed D Greenberger (New York: The New York Academy of Sciences) pp. 304-316

Wheeler J A and Zurek W H 1983 eds *Quantum Theory and Measurement* (Princeton: Princeton University Press)

Wigner E P 1963 *Am J Phys* **31** 6

Wooters W K 1981 *Phys Rev D* **23** 357

Wooters W K and Zurek W H 1979 *Phys Rev D* **19** 473

Zeilinger A 1986 *Physica* **137B** 235

Zeilinger A, Gähler R, Shull C G, Treimer W and Mampe W 1988 *Rev Mod Phys* **60** 1067

A Consistent Interpretation of Quantum Mechanics

Roland Omnès

Abstract
    Some mostly recent theoretical and mathematical advances can be linked
together to yield a new consistent interpretation of quantum mechanics. It
relies upon a unique and universal interpretative rule of a logical
character which is based upon Griffiths consistent historie. Some new
results in semi-classical physics allow to derive classical physics from
this rule, including its logical aspects, and to prove accordingly the
existence of determinism within the quantum framework. Together with decoherence, this can be used to retrieve the existence of facts, despite the
probabilistic character of the theory. Measurement theory can then be made
entirely deductive. It is accordingly found that wave packet reduction is
a logical property, whereas one can always choose to avoid using it. The
practical consequences of this interpretation are most often in agreement
with the Copenhagen formulation but they can be proved never to give rise
to any logical inconsistency or paradox.

    The last decade has brought quite a few technical and mathematical
advances in the foundations of quantum mechanics : Decoherence has been
found to give at least a partial answer to the old problem of a linear
superposition of macroscopic states (Feynmann, Vernon 1963, Hepp, Lieb
1973, Caldeira, Leggett 1983, Zurek 1981 1982). Semi-classical physics has
been better understood by mastering the evolution of coherent states and
by using the powerful mathematical tools of microlocal analysis (Hepp 1974,
Ginibre, Velo 1979, Hagedorn 1980 1981, Omnès 1989). The distinction
to be made between a macroscopic system and a classically behaving one
(Leggett 1980 1984) has become accordingly much clearer. It was also found
possible to describe an history of a quantum system (Griffiths 1984 1987)
and to eliminate meaningless properties such as would occur if a particle
were to cross only one slit in an interference experiment by consistency

conditions needing no appeal to a measuring device. These Griffiths consistent histories were then used to provide a description of a quantum system by ordinary boolean logic (Omnès 1988).

These various results can be used together to provide a new interpretation of quantum mechanics to be called here the logical interpretation. This name is intended to stress the role logic plays in it, on equal footing with empirical evidence and with the mathematical formalism.

This new approach is best understood if one first identifies the main difficulties of quantum mechanics so as to define the problems to be solved. There are two such basic problems, having to do respectively with the status of common sense and the status of empirical facts in quantum mechanics. The first problem is due to the huge gap separating ordinary physical intuition from the mathematical framework of the theory, which is so abstract with its infinite dimensional complex Hilbert spaces having no direct connection with the most usual aspects of reality. Intuition is however necessary, not only because it provides our first approach to reality but also because it is intimately linked with the use of a language allowing common sense to describe reality, to reason simply about it and to exchange this knowledge among ourselves. Despite the high level of mathematical sophistication in physics, one cannot even state its basic principles without having a first intuitive content for many notions. One cannot compare the theory with empirical reality without resorting to common sense in order to describe what happens in the laboratory. Common sense fits more or less easily with classical physics but it is known sometimes to fail when applied to quantum phenomena. In the Copenhagen interpretation, classical physics was assumed to be strictly true for macroscopic systems, so that common sense could go without questioning at this level. However, its relation with quantum properties was never sufficiently investigated.

The second problem has to do with the existence of empirical facts which one considers as providing the foundations of experimental science but which are however at variance with the intrinsically probabilistic character of the theory. The theory only predicts probabilities so that its predictions can only be checked experimentally by performing long series of trials in order to compare empirical frequencies with theoretical probabilities. However, in order to make sense, this procedure must assume that the result of each individual trial is by itself an undoubtable fact, something that is not itself open to probabilistic doubt, that can be

registered, remembered and kept unreservedly for granted. It seems therefore that the basic assumptions of the theory and its empirical contact
with reality are in danger of being contradictory : Because a measuring
devices is made of atoms, it would appear to be described by quantum
mechanics. As such, it would have only uncertain properties described by
their probabilities, not allowing the kind of certainty one generally
attributes to facts. Once again, the Copenhagen interpretation found an
answer to this difficulty by explicitly excluding the use of quantum
concepts when a macroscopic system is concerned.

As a matter of fact, the Copenhagen interpretation stands upon two
quite different sets of physical principles since it uses quantum rules to
describe a microscopic system or the microscopic properties of a macroscopic
object and classical physics to represent the macroscopic properties of a
measuring device or more generally of any such big object. Some room is
therefore left to intuition, facts and common sense since they are supposed
to belong to the classical realm. Interpretation then consists in making
explicit the relation between the two kinds of principles. It is an
endeavour assuming that they can be reconciled, although this essential
consistency had never been proved up to now but had to rest upon an act of
faith.

The starting point of the logical interpretation is much simpler. It
assumes only one kind of principles, quantum mechanics being considered as
universally valid. It does not assume anything about the classical properties of macroscopic objects since they should appear as the outcome of a
theoretical investigation. Interpretation is then not any more an unusual
enterprise of conciliation but just a standard work of deduction as science
does it commonly.

This kind of deduction is however rather far from trivial since one
must now directly face up the two basic problems already mentioned : how
to derive common sense or, if one prefers, classical logic, from the rules
of quantum mechanics and how to explain the occurence of facts with their
certainty within an essentially probabilistic theory ?

The first problem has clearly something to do with logic. It reveals
that a theory does not only involve a mathematical model of reality but
also another construction stating how logic is applied to it in order to
provide the necessary link with empirical reality. In that view, the wide
validity of common sense is the foremost empirical data science must

take care of. This approach was essentially introduced by Von Neumann, although he unfortunately went along two wrong tracks : the first one was to resort to unconventional forms of logic where the most familiar and best formalized one is enough; the second one came from the superposition of macroscopic states that he tried to solve by an appeal to the conscience of an observer with all its well-known philosophical drawbacks. Griffiths idea of consistent histories for a quantum system provided the necessary framework which could then easily be turned into a complete logical construction.

The second problem, having to do with the existence of facts, was then also easily solved by using the progress made in semi-classical physics and the decoherence effect. It was essential however to realize that classical logic, as dealing with the propositions of classical physics, is extremely good but nevertheless only approximate, just as classical dynamics is very good but approximate when the finite value of Planck's constant is taken into account. Without a foundation in a quantum form of logical background, semi-classical physics would never be able to bypass the obstacle raised by the properties of facts. It is in fact most probable that the Copenhagen approach is basically wrong because quantum physics and classical physics cannot be completely but only approximately reconciled, although sufficiently so for most practical purpose.

Despite their very different settings, the Copenhagen interpretation and the logical one are found to agree in most cases of practical interest. This result is very striking since the deductive line of reasoning used in the second approach left no room for adapting it to wise previous knowledge However, it is more precise and there are cases where the two interpretations disagree, as for instance in the case of superconducting quantum interference devices where macroscopic systems can show off quantum properties.

1. GENERAL ASSUMPTIONS

The following conventional basic axioms of rules of quantum mechanics will be assumed :

Axiom 1 associates a specific Hilbert space and an algebra of operators with an individual isolated physical system or, more properly, with any theoretical model of this system.

This axiom will be given a very strong weight, namely everything concerning the system, be it its dynamics or its logical description up to common sense statements about it, should be translatable in strict mathematical terms involving the concepts covered by the axiom.

Axiom 2 defines dynamics in the usual way by a Schrödinger equation in terms of an hamiltonian H. The corresponding evolution operator acting in the Hilbert space will be written as $U(t) = \exp(-2\pi iHt/h)$.

Axiom 3 is technical : The Hilbert space describing two non-interacting systems is the tensor product of their Hilbert spaces and the total hamiltonian is the sum of their hamiltonians.

2. VON NEUMANN PREDICATES

The first logical step going from these very abstract rules towards a more intuitive description of the properties of a system is obtained by using the elementary propositions, or predicates, as they were introduced by Von Neumann (1932).

One should first agree about the vocabulary : A self-adjoint operator A will be called as usual an observable (whatever that means in practice) and any real number belonging to the spectrum of A will be called a value of that observable.

Von Neumann considered the propositions expressing for instance that: "the position X of a particle is in volume V" to be called predicates. Here the particle is associated with a given Hilbert space, X is a well-defined observable and V is a part of its spectrum, so that everything is completely defined in the proposition, namely corresponds with a mathematical concept as allowed by the first axiom, except for the little word "is" or, what amounts to the same, for the meaning of the whole predicate itself. Von Neumann proposed to associate a projector

$$E = \int_V |x \rangle\langle x| \, dx$$

with the predicate to give it a meaning in the Hilbert space grammar of Axiom 1. More generally, with any set C belonging to the spectrum of an observable A, one can associate a predicate [A,C] meaning that "the value of the observable A is in the set C" and a corresponding well-defined projector E. The time-referring predicate stating that the value of A is in C at time t can be associated with the projector $E(t) = U^{-1}(t) \, E \, U(t)$ by taking into account the Schrödinger equation. Conversely, any projector E can be used to define a predicate as can be shown by taking A = E, C being the subset $\{1\}$ in the spectrum $\{0,1\}$ of the projector E.

The second important logical notion to be defined is the state of a system. The difficulty here is that one needs to define it at an early stage whereas its proper meaning only comes out when measurement theory has been constructed, i.e. only at the end.

Axiom 4 assumes that the initial state of the system at time zero can be described by a predicate with projector E_0.

This kind of state description can be shown later on to represent correctly a preparation process in many cases but not all of them. When it holds, a state operator ρ will be defined as the quotient E_0/TrE_0. For instance $\rho = E_0 = |\psi_0><\psi_0|$ in the case of a pure state. In general, ρ must be considered as a density matrix. This is due to the fact that the preparation process may involve the whole past history of the system including sometimes all its past interactions before it became isolated. The exact state of the Earth considered as a physical system might for instance easily take us back to the origin of the Universe.

3. HISTORIES

As introduced by Griffiths (1984, 1987), an history of a quantum system S can be considered as a series of conceptual snapshots describing possible properties of the system at different times. It will be found later on that an history becomes a true motion picture in the classical limit when the system is macroscopic.

More precisely, let us choose a few ordered times $0 < t_1 < ... < t_n$, some observables A_1, ..., A_n which are not assumed to commute and some range of values C_1, ..., C_n for each of these observables. A story $[A_1, ..., A_n, C_1, ..., C_n, t_1, ...t_n]$ is a proposition telling us that at each time $(t_j)(j = 1, 2... n)$, A_j has its value in the range C_j. Despite their seemingly restricted character, these histories contain potentially everything that can be stated about the system.

Griffiths proposed to assign a probability to such a story. We shall write it in the form

$$w = Tr(E_n(t_n)...E_1(t_1) \ \rho \ E_1(t_1)...E_n(t_n)) \qquad (1)$$

Griffiths used a slightly different expression and he relied upon the Copenhagen interpretation to justify it. Here Eq.(1) will be postulated with no further justification, except to notice that it is "mathematically natural" when using Feynman path summations because a projector $E_j(t_j)$ is associated with a window through which the paths must go at time t_j. It should be stressed that w is just for the time being a mathematical measure to be associated with the story, having not yet any empirical meaning that might be found by a series of measurements. Quite explicitly, we dont assume that we know right now what a measurement is.

Griffiths noticed that some restrictions must be imposed upon the projectors entering Eq.(1) in order to satisfy the basic axioms of probability theory and particularly the additivity property of the measures for two disjoint sets. To show what that means, it will be enough to consider the simplest case where time takes only two values t_1 and t_2, denoting by E_1 (resp. E_2) the projector associated with a set C_1 (resp. C_2) and by $\overline{E}_1 = I - E_1$ the orthogonal projector. In that case, it can be proved that all the axioms of probability calculus are satisfied by definition (1) if the following consistency condition holds :

$$Tr \left([E_1(t_1), [\rho, \overline{E}_1(t_1)]] E_2(t_2) \right) = 0 \qquad (2)$$

One knows how to write down similar necessary and sufficient consistency conditions insuring that one is dealing with bona fide probabilities in the general case. The essential point is that they are completely explicit in their mathematical expression.

4. LOGICAL STRUCTURE

Griffiths histories will now be used to describe logically a system in both a rigorous and an intuitive way.

Let it first be recalled that what logicians call a logic or more property an interpretation of formal logic consists in the following construction : One defines a field of propositions (a, b, ...) together with four operations or relations among them, giving a meaning to the logical notions "a or b", "a and b", "not a" and "a implies b", this relation being denoted by a =>b or "if a, then b". This is enough to do logic rigorously if some twenty or so formal rules are obeyed by "and, or, not, if...then". This kind of logic is also called boolean.

In the logical interpretation of quantum mechanics, one meets such a kind of logic at three different levels : It enters first the mathematical notions one is using in the usual form of mathematical logic; however this needs not call any particular attention. It also deals with histories to describe the quantum properties of the system. Finally, when one comes to classical physics, the field of propositions consists of statements concerning classical properties and it will then be called simply classical logic.

Probability calculus is intimately linked with logic. One can make this clear by choosing for instance two times t_1 and t_2 and two observables A_1 and A_2. The spectrum σ_1 of A_1 wil be divided into several regions $\{C_{1\alpha}\}$ and similarly for σ_2. An elementary rectangle $C_{1\alpha} \times C_{2\beta}$ in the direct product $\sigma_1 \times \sigma_2$ will be considered as representing a Griffiths history or what a probabilist would call an elementary event. A union of such sets is what a probabilist considers as an event and its verbal expression will be called here a <u>proposition</u> describing some possible properties of the system.

A field of such propositions will be defined by fixing once and for all some instants t_1, t_2, ...t_n as well as some observables A_1,..., A_n and a complete set of histories covering all the possibilities allowed by a well-defined covering of each spectrum by disjoint sets.

As usual in set theory, the logical operators "and, or, not" will be associated with an intersection, a union or the complementation of sets representing a collection of histories, so that these three logical rules are well defined in the given field of propositions.

When a proposition a is associated with a union of two sets a_1, a_2 each one representing a story, its probability will be defined by

$$w(a) = w(a_1) + w(a_2) \tag{3}$$

and so on. When Griffiths consistency conditions are satisfied, these probabilities are uniquely defined and one can introduce as usual the conditional probability for a proposition b, given some proposition a as given by

$$w(b \mid a) = w(a \text{ and } b)/w(a) \tag{4}$$

Then we shall define implication by saying that proposition a implies proposition b (i.e. a =>b) if $w(b \mid a) = 1$. It can be proved that all the axioms of boolean logic are satisfied by these conventions, as long as the consistency conditions are valid. Once again, despite of its seemingly abstract character, such a construction is able to cover in principle everything that can be said about a physical system.

We shall also introduce a very important notion here : we shall say that a implies b up to an error ε if $w(b \mid a) > 1 - \varepsilon$. This kind of error in logic is unavoidable when macroscopic objects are concerned : When saying for instance that the Earth is at some limited distance of the Sun, one must always take into account a small probability ε for the Earth to leave the Sun and go revolving around Sirius by tunnel effect according to quantum mechanics. The idea consists in introducing logical propositions that are almost certain to follow from some other given ones : for instance, the present situation of the Earth can be stated by a simple predicate. It will be found that this implies another position for the Earth on its orbit tomorrow morning, except for uncontrollable quantum effects that are utterly negligible but can never have strictly zero probability.

You will notice that, even after making sure that the consistency conditions are valid, there remain as many descriptions of a system or as many logics as one can proceed in the choice of the times t_j, the observables A_j and the different ranges for their values. This multiplicity of consistent logics is nothing but an explicit expression of the complementarity principle.

The calculations one can perform with these kinds of logic are more or less straightforward and we shall only mention here one remarkable theorem, albeit a rather simple one : Let us assume that two different logics L_1 and L_2 both contain the same two propositions a and b in their fields of propositions. If a $\Rightarrow$ b in L_1, then a $\Rightarrow$ b in L_2. This theorem means that no contradiction can ever occur so that the construction can never meet a paradox, in so far as a paradox is a logical conflict.

One can now introduce a unique and universal rule for the interpretation of quantum mechanics, stating how to describe the properties of a physical system in ordinary terms and how to reason about these properties:

Axiom 5 : Any description of the properties of a system should be framed into propositions belonging to a consistent logic. Any reasoning concerning them should be the result of an implication or a chain of implications.

From there on, when the word "imply" will be used, it will be in the sense of this axiom. The logical construction allows us to give a clearcut meaning to all the reasoning an experimentalist is bound to make about his apparatuses. In practice, it provides an explicit calculus of propositions selecting automatically the propositions making sense and giving the proof of correct reasonings. Two examples will show how this work :

In an interference two-beams experiment, it is possible to introduce the elementary predicates stating that, at some convenient time t_2, a particle is in one or another region of space where the two beams are recombined. The probabilities of these various predicates allow to describe interference. All the predicates corresponding to different regions describe the possible outcomes of the experiment, although one does not know yet how to describe a counting device registering them. These predicates constitute together a consistent logic. It is also possible to define a projector expressing that the particle was along the upper beam at a previous time. However, it turns out that there is no consistent logic containing this last predicate together with the various predicates describing the outcomes of experiment. This means that logic antecedes measurement : There is no need to invoke an actual measurement to discard the proposition according to which the particle followed only one beam as meaningless; logic is enough to dispose of it according to the universal rule of interpretation, since there is no consistent logic allowing such a statement.

The second example will involve an history which is considered as meaningless in the Copenhagen interpretation but which is perfectly allowed here. Let us consider a particle coming out of a source in an isotropic S-state with a rather well defined velocity. This initial state can be described by an initial projector E_0. Another projector E_2 corresponds to the predicate stating that the particle position is in a very small volume δV_2 around a point x_2 at time t_2. Then, it is possible to choose explicitly a time $t_1 < t_2$, to construct a large enough volume V_1 having its center along the way going from the source to the point x_2 so that one can prove the logical implication : "the particle position is in δV_2 at time $t_2 \Rightarrow$ the particle position is in V_1 at time t_1". So, one can prove in this logical framework that the particle went essentially along a straight trajectory. Similar results hold for the momentum. To speak of both position and momentum at the same time is also possible, as will be seen later on, but of course with some due restrictions.

Simple as they are, these two examples show that the universal rule of interpretation is able to select meaningful propositions from meaningless ones and also to provide a rational basis for some common sense statements which were discarded by the Copenhagen interpretation.

5. CLASSICAL LIMIT

What has been called the universal rule of interpretation makes little avail of what Bohr could have also called a universal rule of interpretation, namely that the properties of a macroscopic device should be entirely described by classical physics. In fact, what he really needed from classical physics was not so much classical dynamics as classical logic where a property can be held to be either true or false, with no probabilistic fuzziness.

Bohr's assumption is not so clearcut as it once seemed since Leggett has shown that some macroscopic systems consisting of a superconducting ring having a Josephson weak link can be in a quantum state (1980, 1984). As a consequence, nobody seems to be quite sure anymore of what the Copenhagen interpretation really states in such a case.

The way out of this puzzle and many other ones was found by showing why and when classical physics, i.e. classical dynamics together with classical logic, holds true as a consequence of the universal interpretative rule. This is of course a drastic change of approach as compared with the familiar course of physics since it means that one will try to prove why and when common sense can be applied rather than taking it for granted as a gift of God. In that sense, it is also a scouring attack against philosophical prejudice.

To begin with, one must make explicit what is a proposition in classical physics. One may think for instance of giving the position and the momentum of a system within some specified bounds. Such a statement is naturally associated with a cell C in classical phase space (in that case a rectangular cell). Since motion will deform such a cell, it looks reasonable to associate a classical predicate with a more or less arbitrary cell in phase space. It will also be given a meaning as a quantum predicate if one is able to associate a well-defined projector $E(C)$ in Hilbert space with the classical cell C in phase space.

If one remembers that, in semi-classical approximations, each quantum state counts for a cell with volume h^n, n being the number of degrees of freedom, two conditions should obviously be asked from the cell C :(i) It must be big enough, i.e. its phase space volume should be much larger than h^n. (ii) It should be bulky enough and with a smooth enough boundary to be well tiled by elementary regular cells. This last condition can be made quite precise and, when both conditions are met and the cell is furthermore connected and simply connected, i.e. in one piece with no hole, we shall say that the cell is <u>regular</u>.

Now there is a theorem stating that an approximate projector E(C) can
be associated with such a regular cell C (Hörmander 1979, Omnès 1989).
To be precise,   one can define it in terms of coherent (gaussian)
states $g_{qp}(x)$ with average values (q, p) for their position and momentum,
putting

$$E(C) = \int |g_{qp}><g_{qp}| \ dq \ dp \ h^{-n} \tag{5}$$

It is easily found that the trace of E(C) is the semi-classical average
number N (= volume of C/h^n) of quantum states in C. In fact, E(C) is not
exactly a projector but one can prove that

$$N^{-1}(C) \ Tr| \ E^2(C) - E(C) | = O((h/LP)^{1/2}) \tag{6}$$

where L and P are typical dimensions of C along configuration space and
momentum space directions. The kind of bound on the trace of an absolute
value operator as met in Eq. (6) is exactly what is needed to obtain classical logic from quantum logics. Using E(C) or a true projector near enough
to it, one is therefore able to state a classical property as a quantum
predicate. This kind of theorem relies heavily upon microlocal analysis
(Hörmander 1985) and, as such, it is non-trivial.

One may extend this description of kinematical properties to dynamical properties by giving a quantum logical meaning to the classical history
of a system. To do so, given the hamiltonian H, one must first find out
the Hamilton function h(q,p) to be associated with it. The answer is given
by what is called in microlocal analysis the Weyl symbol of the operator H
In more familiar terms, the relation between H and h(q,p) is exactly the
one occuring between a density matrix ρ and the associated Wigner distribution function (1939, 1950) $f_w(q,p)$.

By the way, an approximate projector associated with a regular cell
C in phase space can be obtained by choosing for its Weyl symbol the
characteristic function of the cell (equal to 1 in C and to zero outside),
suitably smoothed by a regularization process to make it indefinitely
differentiable.

Once the Hamilton function h(q,p) is thus defined, one can write down
the classical Hamilton equations and find out the cell C_t which is the
transform of an initial regular cell C_0 by classical motion during a time
interval t. Of particular interest is the case when C_t is also regular and
one will then say that the hamiltonian (or the motion) is regular for the
cell C_0 during the time interval t. It will be seen that regular systems

are essentially deterministic, wherefrom their great interest.

Since C_0 and C_t are both regular, one can associate with them two approximate projectors E_0 and E_t as given by Eq.(5), satisfying condition (6). If E_0 were treated like a state operator, it would evolve according to quantum dynamics to become after a time t the operator

$$E_0(t) = U(t) E_0 U^{-1}(t) \qquad (7)$$

Another useful theorem, coming from previous results inaugurated by Hepp (1974) which were further developped by Ginibre and Velo (1979) and Hagedorn (1980, 1981), is the following one (Omnès unpublished) : For a regular system, one has

$$N^{-1}(C_0) \ Tr \, |E_0(t) - E_t| = O(\varepsilon) \qquad (8)$$

Here ε is a small number depending upon C_0, $h(q,p)$ and t, expressing both the effect of classical motion and wave packet expansion. In a nutshell, this theorem tells us that quantum dynamics logically coincides with classical dynamics, up to an error of order ε , at least when regular systems are considered.

This theorem can be used to prove several results concerning the classical behavior of a regular system. Considering several times $0 < t_1 < \ldots$ $< t_n$, and an initial regular cell C_0 becoming, successively via classical motion, the regular cells C_1, ..., C_n, one can use the projectors associated with these cells and their complements to build up quantum propositions. One can then use Eq.(8) to prove that the quantum logic containing all these predicates is consistent. Furthermore, if one denotes by $[C_j, t_j]$ the proposition stating that the system is in the cell C_j at time t_j (it is formulated by stating that the projector $E(C_j)$ has the value 1 at time t_i), one can prove the implications

$$[C_j, \ t_j] \Rightarrow [C_k, \ t_k] \ , \qquad (9)$$

whatever the couple (j, k) in the set (1, ..., n). This implication is valid up to an error ε ,ε being controlled by the characteristic of the cells and the time t_n as explained above.

Eq. (9) has far-reaching consequences. It tells us that classical logic, when expressing the consequences of classical dynamics for a regular system and regular cells, is valid. Of course, it is only valid up to a possible error ε as shown by the example of the Earth leaving the Sun or of a car getting out of a parking lot by a tunnel effect. This kind of probability is essentially the meaning of the number ε and its value is specific for each special case to be considered.

Furthermore, the implications (9) entail that the properties of a regular system satisfy determinism, at least approximately, since the situation at some time t_j implies the situation at a later time t_k. Such a system can also keep a record or a memory since the situation at a time t_j implies the situation at an earlier time t_k. It will be convenient to call such a chain of mutually implying classical propositions a potential fact. This name is used because determinism and recording are essential character-isitics of ordinary facts but one should however not forget that at the present stage the theory is still only just talk-talk-talk with no contact with experiment, wherefrom the term "potential" meaning an imaginary possi-bility.

Since Hagedorn has shown that wave packet spreading is mainly control-led by quantities known from classical dynamics (1981), the property of regularity can in principle be checked completely within classical dynamics. An obvious counter-example of a system not behaving regularly is provided by a superconducting quantum interference device in a quantum situation as described by Leggett (1980, 1984) and investigated by several experimenta-lists (Tesche 1986, Prance et al 1983, 1985, 1986). Another example is given by a chaotic flow after a time t large enough to allow a strong dis-torsion of cells by mixing.

6. DECOHERENCE

The phenomenon of decoherence is by now well known. I shall however recall them briefly for the sake of completeness : Consider for instance a real pendulum, i.e. a ball hanging on a wire. The position of the ball center-of-mass can be characterized by an angle Θ . This angle is a typical example of a collective coordinate. The other coordinates describing the atoms and the electrons in the ball and the wire are the microscopic co-ordinates. Their number N is very large and they are collectively called the environment.

One may start from an initial situation where Θ is given and the velocity is zero. More properly, this can be achieved by using a gaussian state $|\Theta>$ realizing these conditions on average. It may be convenient to assume that the ball and the wire are initially at zero temperature so that the environment is in its ground state $|0>$. So, the complete description of this initial state is given by

$$|\Theta) = |\Theta> | 0 > \tag{10}$$

Naïvely, one would say that the motion of the pendulum will generate deformations of the wire and therefore elastic waves or phonons leading to dissipation. If one compares two clearly different initial situations $|\Theta_1)$ and $|\Theta_2)$, the amount of dissipation in each case after the same time interval will be different so that the corresponding states of the environment will become practically orthogonal as soon as dissipation takes place.

Consider now the initial state

$$|\Psi) = a_1 \; |\Theta_1 >| 0> + \; a_2 \; |\Theta_2> \; |0>$$

and the density operator $\rho = |\Psi><\Psi|$. A collective density matrix ρ_c, describing only the collective coordinate, can be defined as the partial trace of ρ over the environment. Putting $|\psi>= a_1 \; |\Theta_1> + a_2|\Theta_2>$, which is a state of the collective degrees of freedom alone, one finds easily that

$$\rho_c(0) = (a_1|\Theta_1> + a_2|\Theta_2>)(a_1^\star <\Theta_1| + a_2^\star <\Theta_2|) \tag{11}$$

On the other hand, the orthogonality of environmental states noted previously gives, once some dissipation has taken place

$$\rho_c(t) = |a_1|^2 |\Theta'_1><\Theta'_1 + |a_2|^2 \; |\Theta'_2>< \Theta'_2| , \tag{12}$$

the state $\Theta'_1>$ being related to the initial state $|\Theta_1>$ in a way exhibiting motion and damping which needs not interest us here. The essential point is the diagonal form of $\rho_c(t)$ showing the disappearance of phase relations between the two states or what is called decoherence (Zurek 1981, 1982). It shows that the corresponding potential facts are well separated, distinct and mutually exclusive.

As well known, these naive arguments can be replaced by serious proofs (Feynman, Vernon 1963; Hepp, Lieb 1973; Caldeira, Leggett 1983; Zurek 1981, 1982) upon which we shall not elaborate, except for a significant remark.

It has been objected that decoherence does not provide a final proof of fact separation for two different reasons (Zurek 1981, 1982) :

(i) When the collective system is an harmonic oscillator and the environment consists of a bath of harmonic oscillators linearly coupled to it, one can prove the existence of Poincaré recurrences, meaning that the macroscopic pendulum may come back quite near to its initial non-separated situation after a very long time.

(ii) One might try to use a powerful measuring apparatus to detect a microscopic state of the environment which is described by $\rho(t)$, and not by $\rho_c(t)$ and would reveal a non-diagonal matrix element of $\rho(t)$, so that the diagonal form (12) would only be apparent and somewhat irrelevant for the rigorous foundations of quantum mechanics.

These objections are very serious pitfalls for the use of decoherence if one accepts the Copenhagen approach where classical logic is taken as something absolute. Then one can in principle wait long enough to see a Poincaré recurrence or to see the supermeasuring apparatus give at last a non-trivial result. On the contrary, one can easily get rid of the objection by taking into account the status of classical logic as being only an approximation in the logical interpretation. One can easily compute the probability for the occurence of a Poincaré recurrence or for a super-measuring device to give a non-trivial result (i.e. to measure something). These probabilities turn out to be much smaller than the limit of validity of classical logic itself and, as such, it makes no sense to take them into account. Were one to observe them, one would have to attribute them to their most probable cause, namely an exception to the rules of classical physics by the measuring device, and not to an actual success of the experiment.

7. TRUTH, FALSEHOOD AND PERPLEXITY

There is one thing this theory does not explain : How a unique result of an individual measurement is selected among all the possible outcomes. This common sense unicity of facts is far from trivial when no foreign information is introduced into the theory : why there should be a unique fact is far from obvious and my own opinion on the subject is still wavering.

One should surely discard hidden classical variables to provide what philosophers would call an efficient cause because it would be such a pity to reintroduce absolute classical logic at a submicroscopic level when one has been able to get rid of its main troubles at the macroscopic level.

One can however notice a few interesting properties of the present construction. To begin with, it should be stressed that the logical structure of quantum mechanics, as defined by the probability (1) for an history, is not time-reversal invariant. This is the main difference between our choice of a probability and the one that was used by Griffiths.

On the other hand, the theorems of measurement theory to be given in next section rely necessarily upon this choice.

Since facts can be recorded, one can imagine a simple world where all the facts belonging to the past would be recorded. In that case, for any given time t, one can split the flow of time into two separate parts : a past where all the facts are uniquely stated and a future where they remain potential and all their different possible outcomes are allowed. Let us stress that this dichotomy of time into past and future is not a common-place triviality just expressing what common sense shows us but a mathematical property of the logical construction or, once again, a proof of this common sense. As a result, the theory is not able to provide a cause for the unique occurence of a fact but it is able to make place for this uniqueness. Maybe there is no such efficient cause after all and the theory just describes what really is.

To end up with these questions, one can follow Heisenberg's convention by calling true an _actual_ fact (i.e. a unique recorded past fact as opposed to a potential one). However, one may go further by relying upon the non-contradiction theorem mentioned at the beginning and consider a statement as being _reliable_ when it is the logical consequence of a fact. For instance, when I see as a fact the track of a particle in a bubble chamber, I can assert reliably that it came essentially along a straight line before being detected. This is a simple instance where the somewhat formal present theory is nearer to common sense than the Copenhagen interpretation.

Finally, there are propositions which are logically equivalent to a fact : they imply a fact and are implied by it. This is the case for instance for the result of a measurement which is logically equivalent to the data registered by the measuring device. Such propositions strongly related to a fact can also be called true. This is an agreement with the use of the word "true" as it is defined in logic.

8. MEASUREMENT THEORY

Measurement theory now becomes a mere exercize (Omnès 1988). To be specific, we shall only discuss here the measurement of an observable A associated with a physical system Q, all its eigenvalues $\{a_n\}$ being non-degenerate and discrete and the measurement being of the so-called first kind, i.e. preserving this eigenvalue. There is no particular difficulty in extending these results to more general cases.

A measuring apparatus M will be used to measure the observable A. It will be convenient to consider that a collective variable B of M is used to register the measurement data (representing for instance the position of a pointer). In the semi-classical theory of facts previously mentioned, it is possible to take care of friction and damping thus allowing to consider as a data the final position of a pointer on a dial, or its digital recording by a counter. In that last case, the observable B registering the data can only take some well-defined values b_0, b_1, ..., b_n, ... which are the experimental data, after some damping due to an irreversible interaction with the environment lasting a time δ. Initially, the measuring device is assumed to be in a neutral position where the observable B has the neutral value b_0. It should be stressed that the measuring device is treated here by quantum mechanics but that nevertheless the data are treated like facts since this is consistent with quantum mechanics.

It will be assumed that the two physical systems Q and M are initially non-interacting and that, because of some wave-packet overlapping or the triggering of the measuring device, they begin to interact at a time t_0 and do not interact any more after a time $t_1 = t_0 + \delta$ where the data has been recorded by the apparatus M.

M can be assumed to be a perfect measuring device of the first kind for the observable A. This property can be made explicit by introducing the evolution operator $S = U(t_0, t_1)$ for the Q + M system during the time interval where measurement is taking place : It will be assumed that a state $S|a_n>_Q \otimes |b_0 r>_M$, expressing the end-result of the measurement interaction when the initial state of Q is the eigenstate $|a_n>$ of A and the initial state of M is characterized by the neutral initial marking b_0 together with some degeneracy indices r, is a linear superposition of some states $|a_q>_Q |b_m, r'>_M$, where q = m = n. The equality q = n corresponds to a measurement of the first kind and the equality n = m tells us that the measurement is perfect so that the final data represents correctly the initial value of the measured observable. This semi-diagonality of the S-matrix is the only ingredient one needs to define completely measurement theory.

Now the logical game consists in introducing many predicates describing the experiment together with their associated projectors : some of these predicates may describe the history of Q before measurement, some others the history of Q after measurement, one necessary special predicate states the initial value b_0, other ones mentioning the possible outcomes for the final data b_n; finally there are the predicates describing the results of the experiment : some of them express the possible values of A at time t_0 and others at time t_1. One must also introduce the negation of these predicates to obtain a field of propositions for the measurement process forming altogether a logic L.

It is worth stressing that the logical approach forces us to carefully distinguish between the data issuing from an experiment, as shown off for instance by a registering counter, and the result of the experiment itself which expresses the value of the measured observable A.

The first question one has to decide is wether or not the logic L is consistent. To answer it, it is convenient to introduce two other logics L_1 and L_2 referring only to the measured system Q : L_1 describes the history of Q before measurement, including the result $A = a_n$ and its negation at time t_0. L_2 is a logic describing the history of Q after measurement, once the system Q is again non-interacting. It begins by an initial state described by the projector $E_0 = |a_n><a_n|$ at time t_1 and then tells the story of Q after measurement as it was suppposed to be told in L.

One can then prove an important theorem stating that the logic L is consistent if and only if L_1 and L_2 are resepctively consistent.

The occurence of the initial predicate E_0 in the logic L_2 has obviously something to do with wave-packet reduction which is the main theme of this colloquium. It enters the theorem just mentioned because of the semi-diagonality of the S operator, i.e. because one is dealing with a measurement of the first kind.

The precise meaning of wave packet reduction is quite interesting in the present framework. It tells us that it is possible to describe the history of Q after measurement has taken place, when this system has become once again isolated. However, taking the existence of a fact as expressed by the data $B = b_n$ into account forces us to use E_0 as the initial preparation predicate. The basic nature of wave packet reduction is then revealed to be what logicians call in their language a modus ponens: you use for instance a modus ponens when you apply a theorem while forgetting how you proved it and discarding as useless the chains of logical implications contained in the proof of the theorem. A modus ponens is generally a rule for cutting short a long series of logical chains and replacing it by a shorter proposition. It never adds anything deep to the logical construction; as a matter of fact, it does not add anything although its use might be sometimes extremely convenient; in principle, it is always possible to avoid using it.

In the present case, it means that one can discard everything concerning the past history of Q before the measurement as well as the whole history of the measuring device, although keeping the data fact B = b_n into account. Then one can start afresh and tell the story of Q after measurement as an isolated system. The price to be paid for throwing out the past history of Q and the past and continuing history of M consists in the compulsory choice of the initial predicate E_0. However, one can always avoid to use wave packet reduction. In that case, the price to be paid is to use a logic keeping track of the past history of Q as well as of all systems having interacted with it at one time or another. Schematically, it means that a choice is left possible between using wave packet reduction in a cartesian splitting of knowledge or following the whole history of the universe in a grandiose or holistic manner.

Let us now come back to the other theorems of measurement theory. Knowing that the overall logic L is consistent, one can try to prove some implications holding in it. The most interesting one is the following :

$$[B = b_n, \ t_1] \Rightarrow [A = a_n, \ t_0] \tag{13}$$

or, in words : the result $A = a_n$ of the measurement is a logical consequence of the data $B = b_n$. The implication also goes in the other direction so that data and result are logically equivalent. This allows us to say that the result of the experiment is true as a consequence of a fact. This statement is insensitive to the ambiguities brought about by complementarity. The nature of this relation between data and result was left in shadows by the Copenhagen interpretation and this was the main origin of many difficulties such as the ones occuring in the EPR paradox.

Another theorem tells us that, under some trivial restrictions, one can perform once again a measurement of the observable A after a first measurement giving the result a_n has been done : the second result must also be a_n (repetitivity).

Finally, one can compute the probability for the predicate $[B = b_n, t_1]$ describing the experimental data. Because of the semi-diagonality of the S-matrix, this probability turns out to depend only upon the properties of the Q-system and not at all upon the irrelevant degeneracy indices r which represent the model of the apparatus, its type, its color or its age. This probability is simply given by

$$w_n = <a_n \ |U(t_1) \ \rho_Q \ U^{-1}(t_1)| \ a_n> \ , \tag{14}$$

i.e. by the value for the probability of the result $A = a_n$ usually assumed in quantum mechanics.

Using axiom 3, one can now effectively consider a series of independent experimental trials for the measurement of the same observable A upon many identically prepared copies of the system Q. The result of each trial will be a fact and there is nor anymore a difficulty in giving a meaning to the empirical frequency of a given result. The last theorem just mentioned, together with the law of large numbers tells us that, in the limit of a large number of trials, the empirical frequency must coincide with the theoretical probability.

This last theorem is once again a probabilistic theorem, just as the one expressing the validity of classical logic to describe facts. It tells us that "the probability for the empirical frequency to differ from the theoretical probability is negligibly small when the number of trials becomes large enough".

So, finally, one has recovered the main practical results of the Copenhagen interpretation without having to submit to all its limitations and its paradoxes. The exact evaluation of these results as providing perhaps a way out of the difficulties of quantum mechanics will presumably need some time and much discussion and it would be premature to assert it by now. However, it seems rather clear that the resulting interpretation is objective.

References

Caldeira A.O., Leggett A.J. 1983 Ann. Phys. 149 374, erratum 153 445.
Feynmann R.P., Vernon F.L. 1963 Ann. Phys. 24 118.
Ginibre J., Velo G. 1979 Commun. Math. Phys. 66 37.
Griffiths R. 1984 J. Stat. Phys. 36 219.
Griffiths R. 1987 Am. J. Phys. 55 11.
Hagedorn G. 1980 Comm. Math. Phys. 77 1.
Hagedorn G. 1981 Ann. Phys. 135 58.
Hepp K., Lieb E.H. 1973 Helv. Phys. Acta 46 573.
Hepp K. 1974 Commun. Math. Phys. 35 265.
Hörmander L. 1979 Arkiv för Mat. 17 297.
Hörmander L. 1985 The analysis of differential operators, 4 volumes,
    Springer Berlin.
Leggett A.J. 1980 Progr. Theor. Phys. 69 (suppl) 10.
Leggett A.J. 1984 Phys. Rev. B30 1208.

Omnès R. 1988 J. Stat. Phys. 53 893 933 957.
Omnès R. 1989 J. Stat. Phys. 57 n°1/2.
Prance R.J., Mutton J.E., Prance H., Clark T.D., Widom A., Megaloudis G.
  1983 Helv. Phys. Acta 56 789.
Prance R.J. et al 1985 Phys. Lett. 107A 133.
Prance H. et al 1986 Phys. Lett. 115A 125.
Tesche C.D. 1986 Ann. N.Y. Acad. Sc. 480 36.
Von neumann J. 1932 J. Mathematische Grundlagen der Quantenmechanik,
  Springer Berlin.
Weyl H. 1950 Bull. Amer. Math. Soc. 56 115.
Wigner E.P. 1939 Phys. Rev. 40 149.
Zurek W.H. 1981 Phys. Rev. D24 1516.
Zurek W.H. 1982 Phys. Rev. D26 1862.

The Measurement Process in the Individual Interpretation of Quantum Mechanics

Hans Primas

1. VON NEUMANN'S CONCEPT OF QUANTAL MEASUREMENTS

1.1 Spectral measures as basis of von Neumann's measurement theory

Von Neumann's (1927, 1932) measurement theory is based on an uncritical acceptance of the historical disturbance doctrine of measurements. It is an ineffectual attempt to
relate empirical data to the partial Boolean structure of quantum mechanics. In this paper,
I will argue that the measurement problem of quantum mechanics is soluble if it is
reasonably posed. All difficulties stem from an indiscriminate conceptual approach and
inadequate mathematical formulations.

We start our discussion with the historical Hilbert-space formalism of quantum
mechanics in the codification of von Neumann (1932). In this framework, the so-called
observables represent *Abelian* algebras which are related to *Boolean* classifications. This
connection is based on the fact that every bounded or unbounded selfadjoint operator A
acting on a Hilbert space $\mathcal{H}$ has a unique *spectral representation*

$$A = \int_{\Lambda} \lambda\, E(d\lambda) \quad , \tag{1}$$

where $\Lambda \subset \mathbb{R}$ is the spectrum of A, and $E: \Sigma \to \mathcal{B}(\mathcal{H})$ is a normalized spectral measure on
the σ-field Σ of Borel subsets of Λ. A normalized *spectral measure* $E: \Sigma \to \mathcal{B}(\mathcal{H})$ on an
arbitrary measurable space (Λ, Σ) is a projection-valued set function having the following
properties:

(i) E is nonnegative-definite, that is

$$E(B) \geq 0 \quad \text{for every } B \in \Sigma \quad . \tag{2a}$$

(ii) E is *additive*, that is

$$\text{if } B_1 \cap B_2 = \varnothing \; , \text{ then } E(B_1 \cup B_2) = E(B_1) + E(B_2) \quad . \tag{2b}$$

(iii) E is *continuous*, that is whenever $\{B_n\}$ is an increasing sequence of sets
in Σ whose union is B, then

$$\sup E(B_n) = E(B) \quad . \tag{2c}$$

(iv)   *E* is *normalized,* that is

$$E(\Lambda) = 1 \quad . \tag{2d}$$

(v)   *E* is *multiplicative,* that is

$$E(B_1 \cap B_2) = E(B_1)E(B_2) \quad . \tag{2e}$$

Every spectral measure generates a Boolean projection lattice $\mathcal{P}$ and a commutative algebra C,

$$\mathcal{P} = \{E(B) \mid B \in \Sigma\} \quad , \tag{3}$$
$$C = \{E(B) \mid B \in \Sigma\}'' \quad , \tag{4}$$

where " denotes the double commutant.

1.2   The expectation-value postulate of traditional quantum mechanics

In von Neumann's codification, a minimal operational assessment of the mathematical formalism is given by the so-called *expectation-value postulate* which is of crucial importance for all engineering applications of quantum mechanics.

Von Neumann's expectation-value postulate

Let A be a selfadjoint observable with spectrum Λ, let Σ be the σ-algebra of Borel sets of Λ and E be the spectral measure of A. The probability that a predictive measurement of the first kind of the observable A gives a value lying in the Borel set $B \in \Sigma$ is given by the Kolmogorov probability measure $\mu : \Sigma \to \mathbb{R}$, defined by

$$\mu(B) = \langle \Psi | E(B) | \Psi \rangle \quad , \tag{5}$$

where Ψ is the normalized state vector immediately before the measurement.

Conditioned by a measurement of the first kind, this postulate implies that the observable A can be considered as a conditional random observable on the Kolmogorov probability space (Λ, Σ, μ). The conditional expectation $\mathcal{E}(A)$ of this random variable is given by

$$\mathcal{E}(A) = \int_\Lambda \lambda \, \mu(d\lambda) = \langle \Psi | A | \Psi \rangle \quad . \tag{6}$$

If one adopts the usual frequency interpretation of Kolomogorov's probability theory, then one anticipates that the average of the numerical results of a large number of measurements of the observable A approaches asymptotically the expectation value $\mathcal{E}(A)$. With this stipulation, the expectation-value postulate assigns an empirical meaning to the abstract mathematical formalism. In spite of the associated unrealistic idealizations, the expectation-value postulate is a well-confirmed *working rule* whose success has to be explained by a *fullfledged* theory of the measurement process.

1.3 Descriptions of measurements of the first kind

Measurements which will give the same value when immediately repeated have been introduced for the first time by Johann von Neumann (1927), and are called *measurements of the first kind* (Pauli, 1933). According to a basic tenet of quantum theory, a measurement of the first kind of an observable with a purely discrete nondegenerate spectrum yields as its numerical result one of the eigenvalues of the observable. Von Neumann generalized this postulate and linked measurements of the first kind to projection-valued measures, hence via the spectral theorem to selfadjoint observables. Before we discuss the problematic nature[1] of this connection, let us discuss the mathematical descriptions of measurements of the first kind by means of the simplest example.

Consider a two-level quantum system with the system observable

$$\sigma_3 = |\alpha\rangle\langle\alpha| - |\beta\rangle\langle\beta| \quad , \tag{7}$$

a normalized initial state vector $|\chi_i\rangle = c_\alpha |\alpha\rangle + c_\beta |\beta\rangle$, $|c_\alpha|^2 + |c_\beta|^2 = 1$, and an initial density operator $D_i = |\chi_i\rangle\langle\chi_i|$. A «measurement of the first kind» should actualize the observable σ_3, such that in every single case the measured value is either $+1$ or -1, whereby the value $+1$ occurs with the relative frequency $|c_\alpha|^2$, and the value -1 occurs with the relative frequency $|c_\beta|^2$.

In a statistical ensemble description of the measuring process, the initial state is described by the density operator $D_i = |\chi_i\rangle\langle\chi_i|$. After a measurement of the first kind, the final statistical state of the object system is given by the density operator

$$D_f = |c_\alpha|^2 |\alpha\rangle\langle\alpha| + |c_\beta|^2 |\beta\rangle\langle\beta| \quad , \tag{8}$$

so that in the statistical description the measuring process is described globally by the *linear* map $D_i \rightarrow D_f$. Note that this map alone does *not* allow to say that in the statistical average the event "spin up" happens with the relative frequency $|c_\alpha|^2$, since the space of statistical states is not a simplex, so that the ignorance interpretation is not admissible.

In an individual description, a measurement of the first kind transforms the initial state vector $|\chi_i\rangle$ either into the state vector $|\chi_f\rangle = |\alpha\rangle$ (with probability $|c_\alpha|^2$), or into the state vector $|\chi_f\rangle = |\beta\rangle$ (with probability $|c_\beta|^2$). Accordingly, in the individual description a measurement of the first kind is characterized by a stochastic and *nonlinear* map $|\chi_i\rangle \rightarrow |\chi_f\rangle$.

---

[1]   We restrict our discussion to the widely accepted case of observables having a purely discrete and nondegenerate spectrum. An appropriate formulation of the projection postulate for the case of a degenerate discrete spectrum has been given by Lüders (1951). Von Neumann's remarks about measurements of observables with continuous spectra are untenable, compare Srinivas (1980), Ozawa (1984).

Von Neumann's projection postulate $D_i \twoheadrightarrow D_f$ is a necessary, but *not a sufficient condition for a measurement of the first kind*. The so-called «reduction of the wave packet» $|\chi_i\rangle \twoheadrightarrow |\chi_f\rangle$ implies the projection postulate $D_i \twoheadrightarrow D_f$, but the map $D_i \twoheadrightarrow D_f$ does not imply the map $|\chi_i\rangle \twoheadrightarrow |\chi_f\rangle$. The statistical projection postulate is a reduced description which includes not enough information to describe what is going on during the measuring process. A full description of the measurement process is feasible in terms of *algebraic quantum mechanics*. Let $\mathcal{M}_{obj}$ be the W*-algebra of observables of the object system and $\mathcal{M}_{meas}$ the W*-algebra of observables of the measuring system, so that the W*-algebra of observables of the combined system is given by

$$\mathcal{M} = \mathcal{M}_{obj} \otimes \mathcal{M}_{meas} . \tag{9}$$

The initial state ρ_i and the final state are positive and normalized elements of the predual

$$\mathcal{M}_* = (\mathcal{M}_{obj})_* \otimes (\mathcal{M}_{meas})_* \tag{10}$$

of the W*-algebra $\mathcal{M}$. The joint initial state is given by

$$\rho_i = \omega \otimes \Omega \;\; , \;\; \omega \in (\mathcal{M}_{obj})_* \;\; , \;\; \Omega \in (\mathcal{M}_{meas})_* \;\; , \tag{11}$$

while the final joint state is given by

$$\rho_f = |c_\alpha|^2 \, \omega_\alpha \otimes \Omega_\alpha + |c_\beta|^2 \, \omega_\beta \otimes \Omega_\beta \tag{12}$$

where Ω_α and Ω_β are *disjoint* states of the measuring instrument. In an individual interpretation the states ω, ω_α and ω_β of the object system, and the states Ω, Ω_α and Ω_β of the measuring system are pure states. A necessary and sufficient condition that von Neumann's projection postulate is equivalent to the «reduction of the wave packet» is that the two apparatus states Ω_α and Ω_β are *disjoint*[1]. In the special case where an irreducible Hilbert-space representation of the algebra $\mathcal{A}$ exists, the state vectors $|\Psi_\alpha\rangle = |\alpha\rangle \otimes |\Xi_\alpha\rangle$ and $|\Psi_\beta\rangle = |\beta\rangle \otimes |\Xi_\beta\rangle$ corresponding to disjoint pure states $\omega_\alpha \otimes \Omega_\alpha$ and $\omega_\beta \otimes \Omega_\beta$ lie in different sectors (i.e. they are separated by a superselection rule).

The measurement problem is *not* – as often asserted – how a pure state can be transformed into a mixed state. In the statistical description this would be a trivial task (most dynamical linear semigroups and their Hamiltonian dilations can describe such a map), while in the individual description there are no mixed states at all. What is to be explained is in the statistical description the emergence of *disjoint* final statistical states, or equivalently, in the individual description the emergence of the *nonlinear* und *stochastic* map which transforms the initial state vector into *one* of the possible final state vectors. If we take this delineation as the proper formulation of the "collapse of the wave packet", then this problem can be tackled with the tools of modern mathematical physics.

---

[1] Two pure states ω and ρ on a W*-algebra $\mathcal{M}$ are called disjoint if there exists a classical observable C in the center of $\mathcal{M}$ such that $\omega(C) \neq \rho(C)$. For more details, compare Bratteli and Robinson (1979).

Ludwig (1985b, p.329) claims that "the entire problem of the «collapse of the wave packet» is not a real problem but is a problem which arises from the addition of unnecessary ideas", but he just assumes and does not prove that the final states are *disjoint* ones. Yet, exactly the additional unnecessary ideas (namely the individual interpretation) give us the simplest tools to prove rigorously the emergence of a superselection rule associated with every measurement process.

2. REPEATABLE MEASUREMENTS DO NOT EXIST

2.1 *Quantal observables never can be measured exactly*

According to von Neumann (1932, chapt.III.3), an observable can be measured *exactly* if it possesses a pure discrete spectrum. *This statement is profoundly mistaken.* Even for spin $\frac{1}{2}$ there exist no dispersion-free yes-no experiments. A process $|\chi_i\rangle \rightarrow |\chi_f\rangle$ with $|\chi_f\rangle = |\alpha\rangle$ or $|\chi_f\rangle = |\beta\rangle$ refers to a singular situation which is experimentally neither repeatable nor verifiable. *In quantum mechanics, the experimentally relevant sample space is always uncountable,* even for the simplest case of a two-dimensional Hilbert space.

Every laboratory experiment is, in its last stages, describable without quantum mechanics in terms of a Boolean language. All measurements can always be executed in a digital manner and can be characterized by a *finite* Boolean algebra. If the theoretical sample space of an analog output signal is uncountable, one is forced to use some classification method together with a statistical decision procedure. Let $\{B_i \mid i = 1,...,n\}$ be a finite, experimentally feasible partition of Ω, i.e.

$$B_j \cap B_k = \varnothing \quad \text{for } j \neq k \quad , \quad B_1 \cup B_2 \cup \cdots \cup B_n = \Omega \quad . \tag{13}$$

If the output signal lies in the set B_i, we say the event B_i has occurred. This classification of observation data into disjoint groups brings about a *Boolean classification*. The rule that experiments have to be *reproducible* does not refer to individual experimental data but to *equivalence classes* of data. Which data we condsider as equivalent depends on the context. Philosophers often like to regard measurements as objective, and forget the long learning process about how to carry out measurements and reject spuriae. *A fact is only a fact if we declare it as a fact.* In order to get "empirical facts", decision rules are inevitable. That is, *von Neumann's repeatability axiom refers to the equivalence classes and decision procedures of experimental science, and should not be connected directly with fundamental first principles that are supposed to describe material reality in itself.*

2.2 *How to describe a measurement for a two-level quantum system*

As an example, consider again a measurement of the observable σ_3 of a spin-$\frac{1}{2}$ system. According to von Neumann's projection postulate, we expect that the final state vector is given either by $|\alpha\rangle$ or by $|\beta\rangle$. In order to get an experimentally verifiable statement, we parametrize the family of all pure states of the Hilbert space $\mathbb{C}^2$. The set of all pure states in $\mathbb{C}^2$ is isomorphic to the unit sphere in three dimensions, so that a convenient parametrization is given by the Cayley-representation with the Euler angles ϑ and φ, $0 \le \vartheta \le \pi$, $0 \le \varphi < 2\pi$. We define a normalized state vector $|\vartheta, \varphi\rangle$ by

$$|\vartheta, \varphi\rangle := e^{+i\varphi\sigma_3/2} e^{-i\vartheta\sigma_2/2} |\alpha\rangle = e^{+i\varphi/2} \cos(\vartheta/2) |\alpha\rangle + e^{-i\varphi/2} \sin(\vartheta/2) |\beta\rangle . \quad (14)$$

These pure states form an overcomplete set and generate a probability operator measure (i.e. a normalized positive operator-valued measure) $P : \Sigma \twoheadrightarrow \mathcal{B}(\mathbb{C}^2)$,

$$P(B) = \int_B |\vartheta, \varphi\rangle\langle\vartheta, \varphi| \, \sin(\vartheta) \, d\vartheta \, d\varphi \quad , \quad B \in \Sigma , \quad (15)$$

where Σ is the σ-field of Borel sets of the unit sphere. If the measuring apparatus is axial symmetric, the relevant probability operator measure for a σ_3-measurement is given by

$$P([\vartheta_1, \vartheta_2]) = \int_0^{2\pi} \int_{\vartheta_1}^{\vartheta_2} |\vartheta, \varphi\rangle\langle\vartheta, \varphi| \, \sin(\vartheta) \, d\vartheta \, d\varphi \quad , \quad 0 \le \vartheta_1, \vartheta_2 \le \pi . \quad (16)$$

All experimental facts depend in one way or the other on statistical decision tests which imply a partitioning of the sample space. For example, we may consider the following three hypotheses

(i)   the observation comes from the population $0 \le \vartheta \le \varepsilon$,

(ii) the observation comes from the population $\pi - \varepsilon \le \vartheta \le \pi$,

(iii) the observation comes from the population $\varepsilon < \vartheta < \pi - \varepsilon$.

The experimentalist must have sufficient information about the error probability of his measuring instrument, so that he can make statistical tests (Rényi, 1966). For an appropriate choice of ε (where for a well-constructed instrument $\varepsilon \ll 1$) he may, for example, adopt the following decision rule: if the measured value ϑ_0 is in the interval $[0, \varepsilon]$, then the measured value of σ_3 is taken to be $+1$; if it is in the interval $[\pi - \varepsilon, \pi]$, then the measured value of σ_3 is taken to be -1; if it is in the interval $[\varepsilon, \pi - \varepsilon]$, then it is assumed that an error has occurred. Every experimenter will check whether his experimental arrangement leads to reproducible results or not. He will decide that his experiments are reproducible if the frequency of events in the interval $\varepsilon < \vartheta < \pi - \varepsilon$ is inconsequential. From this point of view, von Neumann's repeatability axiom refers to the Boolean classification of the final statistical decision method necessary to arrive at "experimental facts", and not to the Boolean structure of the projection-valued measures associated with the so-called observables in terms of which the first principles of quantum mechanics are formulated.

2.3   *Probability operator measures as basis of a realistic measurement theory*

A *probability operator measure* on an arbitrary measurable space (Λ, Σ) is a normalized positive operator-valued measure, i.e. a map $F: \Sigma \twoheadrightarrow \mathcal{B}(\mathcal{H})$ having the properties (2a), (2b), (2c) and (2d) *but not* (2e). For a probability operator measure to be a spectral measure it is necessary and sufficient that it is multiplicative in the sense of eq.(2e). For every Borel set $B \in \Sigma$, the operator $F(B)$ of a probability operator measure is selfadjoint and non-negative definite, but in contrast to the operators of a spectral measure not necessarily idempotent. Every pair (F, ρ) of a probability operator measure $F: \Sigma \rightarrowtail \mathcal{B}(\mathcal{H})$ and a normal state ρ on the W*-algebra $\mathcal{B}(\mathcal{H})$ generates a probability measure $\mu: \Sigma \rightarrowtail \mathbb{R}$

$$\mu(B) = \rho\{F(B)\} \quad , \quad B \in \Sigma \quad . \tag{17}$$

Experimentally measurable quantities are represented by probability operator measures $F: \Sigma \rightarrowtail \mathcal{B}(\mathcal{H})$ from a Boolean σ-algebra Σ of measurement outcomes into the set of effects $F(B)$ which are non-negative operators with spectrum within the interval [0,1]. That is, probability operator measures – and not spectral measures – are the appropriate tools to describe measurements both for observables with purely discrete spectrum and with continuous spectrum[1]. An operationally meaningful generalization of von Neumann's expectation-value postulate is straightforward:

Operational expectation-value postulate
The probability that a predictive measurement of a generalized observable associated with a probability operator measure $F: \Sigma \rightarrowtail \mathcal{B}(\mathcal{H})$ gives a value lying in the Borel set $B \in \Sigma$ is given by the Kolmogorov probability measure $\mu :$ $\Sigma \to \mathbb{R}$, defined by

$$\mu(B) = \rho\{F(B)\} \quad , \tag{18}$$

where ρ is the statistical state immediately before the measurement.

Philosophers should know by now that the so-called "problems" of joint measurements of noncommuting observables[2], of sequential quantal measurements[3], of the position measurement for photons[4] are solved since long in terms of probability operator measures, not only on the basis of an operational axiomatic foundation[5], but also in the best engineering tradition in terms of optimal estimation and detection theory[6].

2.4 A proper formulation of the expectation-value postulate does not yet solve the "problem of the reduction of the wave packet"

Measurements of the first kind do not exist as laboratory measurements. *No measurement is instantaneous, no measurement is exactly repeatable, hence no measurement obeys strictly von Neumann's formulation of the expectation-value postulate.* However, the fact that von Neumann's projection postulate is utterly unrealistic does not mean that a physically reasonable rephrasing in terms of probability operator measures would already solve the crucial question in the theory of the measurement processes, namely

---

1   Compare: Davies and Lewis (1970), Davies (1970, 1976), Srinivas (1980), Ozawa (1984).
2   Compare: Arthurs and Kelly (1965), Gordon and Louisell (1966), She and Heffner (1966), Holevo (1973 a, b, 1974)
3   Compare: Benioff (1972 a, b, c, 1973).
4   Compare: Jauch (1967), Amrein (1969), Ali (1974), Kraus (1977).
5   Compare: Ludwig (1970, 1983, 1985 a, 1987).
6   Compare: Helstrom (1976), Holevo (1978, 1982).

how the measurement process can be reconciled with the deterministic Schrödinger equation and the quantum-mechanical superposition principle. Nevertheless, a correct descriptive formulation of the measurement operation is imperative for a proper dynamical theory of the measurement process.

Furthermore, an understanding of the expectation-value postulate relies on our previous intuitive notions about physical measurements which are not codified as first principles of quantum theory. From a conceptual point of view, much more needs to be said, for example about the nature of time (e.g. the earlier-later relationship), about the status of facts (does quantum mechanics admit factual descriptions?), or about the feasibility of making experiments (does a theory with an automorphic time evolution allow observers having free will?). If the world external to the quantum object is taken to be given phenomenologically, this vague intuitive formulation of the expectation-value postulate is sufficient to compare quantum-mechanical predictions with experimental results. For a fundamental theoretical approach, we cannot avoid to take the bull by the horns.

3. INTRINSIC AND OPERATIONAL DESCRIPTIONS

3.1 On the difference between intrinsic and operational descriptions

All fundamental first principles we know refer to *strictly closed systems*, whose dynamics are given by one-parameter groups of automorphisms. Since strictly closed systems cannot be observed from outside, such descriptions are not operational and they should not be applied thoughtlessly to open systems.

Abner Shimony (1984) distinguishes between intrinsic and operational theories. An *intrinsic theory* characterizes the intrinsic properties and the states of a physical system independently of any other physical systems or contexts. By contrast, an *operational theory* makes explicit reference to other physical systems and appropriate contexts. It typically characterizes the modes of reaction of the object system in terms of test procedures which must be described in terms of external systems. Both approaches represent to some extent an ideology, a precept of what physics should be like.

Operational and intrinsic formulations are by no means equivalent, there is not even a smooth blend of first principles and applications. A very far developed operational codification of quantum mechanics has been worked out by Günther Ludwig (1983, 1985a, 1987) and his Marburg school. The emphasis of this approach lies on observed data and on an objective description of the experimental tools. It is claimed that all concepts must be constructed from an experimental point of view. According to Ludwig, a physical theory like quantum mechanics has to be interpreted from outside in terms of craftmen's trade. Such an operational approach is appropriate for all problems in *engineering* science.

Engineers and philosophically minded scientists may have different opinions of the aim of scientific theories. A consequent *homo faber* may reject any assertion without empirical content, but the claim that science is *nothing else* than an instrument for organizing and predicting human experience is sheer nonsense. The motivation for experimenting is the desire to learn something about the external reality, a reality which is taken to be independent of our thinking and acting. Whatever positivists may say, many scientists stubbornly hold that there is a real external word, and that the propositions of a *fundamental* scientific theory should be either true or false, *and that what makes them true or false should be something external of our mind*. Of course, a belief in external reality is neither provable nor refutable, it is a statement of faith, which is widely accepted in our everyday life. Realism is a truly *metaphysical regulative principle*, yet science cannot be divorced from metaphysics without becoming rootless and loosing its meaning. The problem of realism in science is the question of the status of things while they are *not observed*, the problem of measurement is the question *which features* of the observed data are due to our mental organization.

3.2   Endo- and exophysical descriptions

A helpful distinction has been made by David Finkelstein (1988) and Otto Rössler (1987) by introducing the notions of endophysics and exophysics. A physical system without an external observer is called an *endosystem*. If the observer is external, we speak of an *exophysical description* of the system under investigation. The world of the observer with his communication tools is called the *exosystem*.

Both endo- and exophysical descriptions have a proper and important place in science, but they must not be confused. The first principles of a universally valid theory refer to endophysics, yet they are not sufficient for a characterization of an exosystem. Therefore, as stressed by Rössler, *endophysics is different from exophysics*. In quantum theory, even the mathematical formalism for quantum endophysics is different from the formalism for quantum exophysics. Many of the conceptual difficulties and alleged paradoxes of quantum mechanics are due to the failure to distinguish properly between endophysical and exophysical descriptions. The first principles of a universally valid theory refer to *strictly closed* systems, they are not sufficient for a characterization of exosystems.

All fundamental first principles we know belong to deterministic endophysics. Experimental physics belongs to exophysics. Every exophysical description requires a division of the holistic quantum world. Yet, the endoworld does not present itself already divided. *We have to divide it.*

3.3   On the validity of quantum mechanics

We adopt the *working hypothesis* that in a proper codification quantum mechanics has universal and unlimited validity at the level of atoms, molecules and non-cosmological macroscopic bodies. Subatomic, genuinely Lorentz-relativistic phenomena, and problems related to men's free will and consciousness will, however, be excluded.

Quantum mechanics is a much richer theory than initially expected, and there is no indication that it could be inadequate for the description of the behavior of matter in the molecular and non-cosmological macroscopic domain. At present, there are no good reasons to consider a program of "completing" quantum mechanics by some so-called «hidden variables» that should more comprehensively specify the state of the system, or any other ad hoc modification of quantum mechanics. Recent proposals to test whether nonlinear corrections to the framework of quantum mechanics are necessary (Shimony, 1979, Weinberg, 1989), fail to make distinctions between individual and statistical, and between endophysical and exophysical descriptions. Proper statistical descriptions are always governed by *linear* evolution equations, but this linearity has nothing to do with the quantum-mechanical superposition principle. On the other hand, even if the quantum-mechanical superposition principle is strictly valid on the endophysical level, it is possible (and in fact generically so) that an individual reduced exophysical description of a quantum object is rigorously governed by a *nonlinear* Schrödinger equation[1].

Endophysics – and not exophysics – is the domain where one can hope to find universally valid first principles. In this connection, «universally valid» does not mean that in its domain of validity the theory can describe everything at the same instant, but only that it can describe any selected partial object. This state of affairs has nothing to do with quantum mechanics but is a *logical* necessity. In order to avoid paradoxes of self-reference, any operationalization of an endophysical theory must be formulated in another, essentially richer language, the *exophysical metalanguage*[2].

An endophysical codification of quantum mechanics has a time-reflection invariant dynamics and leads to a bidirectional deterministic time evolution for its states, so that it is intrinsically a *nonprobabilistic* theory. That is, *quantum endophysics is deterministic*. This does not imply that it is determinable by an observer. The primary probabilities of quantum mechanics (Pauli, 1954) manifest themselves only in the interaction with classical external systems. Since quantum endophysics is a non-probabilistic theory, its natural referent is a *single system*, and not a statistical ensemble. On the other hand, quantum exophysics is, by its very nature, a *statistical theory* which presupposes the existence of statistical *Boolean* classification devices, hence of *external* classical measuring systems, which have a nonanticipating Boolean memory, and which show an irreversible dissipative behavior.

---

[1]   For a more detailed discussion of the linearity of quantum dynamics, compare Primas (1990 a, b).
[2]   For further details and references, compare Primas (1990 a, in press).

3.4 On the the relations between endo- and exophysics

We presuppose that quantum endophysics is universally valid, that within the endosystem the quantum mechanical superposition principle has unrestricted validity, and that the dynamics is time-inversion invariant and non-probabilistic. On the other hand, exophysical descriptions depend on well-specified contexts; they are as a rule not time-inversion invariant, but are characterized by many spontaneously broken symmetries, and generically governed by nonlinear dynamical equations. *The measurement problem of quantum mechanics is the question of how to derive an exophysical description of a quantum object from quantum endophysics.*

This problem cannot be solved within the framework of von Neumann's (1932) codification of quantum mechanics since it is based on the uniqueness theorem for the canonical commutation relations, hence valid only for strictly isolated systems with only finitely many degrees of freedom. The measurement process cannot be described by a unitary transformation of the Hilbert space vector of the composite system, simply because for a realistic measuring system there exists no unique Hilbert-space representation. Yet, algebraic quantum mechanics – a straightforward generalization without any ad hoc modifications – allows a proper formulation of quantum endophysics as a dynamical C*-system in an individual ontic interpretation (Primas, 1990 a). An exophysical description can be derived therefrom by specifying an appropriate exosystem with the aid of the so-called GNS-construction. This procedure leads to a dynamical W*-system which, as a rule, has many classical observables and shows many spontaneously broken symmetries. The crucial point is that a nontrivial endophysical C*-system (i.e. a system whose algebra of intrinsic observables contains noncompact operators) has very many faithful but *physically inequivalent* W*-representations. There is no universal, unique or distinguished exophysical description of a nontrivial quantum object. *We* have to select a representation which corresponds to the tacit idealizations every experimenter is making.

4. A COMPLETE DESCRIPTION OF THE MEASUREMENT PROCESS REQUIRES AN INDIVIDUAL INTERPRETATION

4.1 What is an interpretation of quantum mechanics?

The word "state" is commonly used in serveral conceptually different senses. First of all, one has to distinguish between individual and statistical descriptions. Both are feasible, but they use different formalisms and conceptually different state concepts (Primas, 1990 a). If one adopts tacitly a statistical ensemble description, one may consider a state vector as providing merely a "state of knowledge", and not an objective description of some aspects of physical reality, so one may claim that "a single system has no state" (Peres, 1974). However, without a precise qualification this proposition is false.

An interpretation always refers to a logically consistent and empirically well-confirmed theoretical formalism. That is, we assume that we have a mathematically rigorous codification of a physical theory (the «formalism»), a minimal interpretation of the theory which allows an operationalization and an empirical verification of the theoretical propositions. We adopt the following definition: *An interpretation of a physical theory is characterized by a set of normative regulative principles which can neither be deduced nor be refused on the basis of the mathematical codification and the minimal interpretation.*

We distinguish between epistemic and ontic interpretations. Epistemic interpretations refer to our knowledge of the properties or modes of reactions of systems, while ontic interpretations refer to the properties of the object system itself, regardless of whether we know them or not, and independently of any perturbations by observing acts. The operationalistic view requires an epistemic interpretation and usually works with a statistical ensemble description. A realistic world view demands an individual ontic interpretation of *quantum endophysics*, it is intrinsically objective but not operational. By a proper choice of the regulative principles, one can get a contextually objective and operational *exophysical* description of the quantum reality.

4.2 What are individual quantum objects?

Sometimes it is claimed that *any* attempt to define a pure state to individual quantum systems leads to paradoxes and inconsistencies (Peres, 1974). The reason for this false statement is a failure to distinguish between the formalisms of individual and statistical descriptions. There is a logically consistent formulation of quantum mechanics of *individual* systems which is compatible with all empirical data. (Primas, 1987; 1990 a,b). A necessary and sufficient condition for the feasibility of such an individual interpretation is that the referents of the theory are *objects*. We distinguish between the concepts «system» and «object». By a system we mean nothing but the referent of a theoretical discussion (specified, for example, by a Hamiltonian), without any ontological commitment. On the other hand, we choose the term *object* as the most general ontological expression for something that *persists when not perceived*, for something having *individuality* and *properties*. We do not restrict the notion of an "object" to concrete things or to entities which are localized in space. Objects may be quite abstract individuals which not necessarily can be isolated from the rest of the world. Objects are entities which retain their identity in the course of time. They may change their actualized properties but they keep their identity. In quantum theory, *we define an object as an open quantum system, interacting with its environment, but which is not EPR-correlated with the environment.* Quantum systems which are not objects are entangled with their environments, they have no individuality and allow only an *incomplete* description in terms of statistical states. Since in quantum theories the set of all statistical states is not a simplex, statistical states have no unique decomposition into extremal states. This fact leads to grave problems for a purely statistical interpretation of quantum mechanics. *A conceptually sound statistical ensemble interpretation has to be based on an individual interpretation.*

4.3 How can we describe individual quantum objects?

Von Neumann's codification of 1932 refers to strictly isolated objects which are a priori of no physical interest whatsoever. That is, within this very restricted formalism nontrivial objects do not exist. But it would be wrong to put the blame on quantum mechanics, it just characterizes one of the limitations of von Neumann's codification. In algebraic quantum mechanics the following result holds: *Every classical system is an object. A nonclassical open system is an object if and only if its environment is classical.* (Raggio, 1981).

It can be proved that there is a one-to-one correspondence between the ontic states of an object and the extremal, normalized positive linear functionals on $\mathcal{A}$ (the so-called «pure states»). Furthermore, it follows that a property represented by an observable $A \in \mathcal{A}$ is actualized at time t if and only if $\rho_t(A^2) = \{\rho_t(A)\}^2$, where $\rho_t \in \mathcal{A}^*$ represents the ontic state at time t.

These mathematical results of algebraic quantum mechanics give the green light for an ontic interpretation. The referents of an ontic interpretation of quantum mechanics are *individual objects*. The *potential properties* are represented by the selfadjoint elements of an appropriate C*-algebra $\mathcal{A}$. The *ontic state* of an object at time t is characterized by the set of all potential properties which are actualized at the instant t. That is, *the potential properties characterize the object, while the actualized properties characterize the state of the object.*

4.4 What is a classical description of a quantum system ?

Since quantum *objects* exist if and only if classical environments exist, it is important to investigate classical quantum systems. *We call a quantum system classical if its algebra of observables is commutative.* Of course, our use of the word «classical» has nothing to do with Planck's constant. Furthermore, whether a system shows a quantal or a classical behavior is independent of the size of the object.

The unfortunate historical identification of the classical/quantal dichotomy with the macroscopic/microscopic dichotomy is responsible for many confusions. Since there is a practically continuous transition between microphysics and macrophysics, the distinction between microphysics and macrophysics has become theoretically obsolete. Nowadays, we know that there are microscopic systems having classical properties and macroscopic systems having quantal properties. The macroscopic nature of most measuring instruments of classical science is entirely irrelevant for the genuine quantum mechanical measurement problem. From an engineering point of view, amplification devices may be desirable, but conceptually they are not necessary. One should not forget that the biological *molecular code* realizes at the molecular level a highly reliable Boolean, irreversible and causal memory which in fact realizes a measuring instrument of microscopic size.

5.　THE DYNAMICS OF MEASUREMENT-TYPE PROCESSES

5.1　*Necessary preconditions for a measurement*

Physical measurements are intended to create facts. Factual descriptions refer to the past, stand and fall with irreversibility, must be expressed in a Boolean language, and have to be registered. Registration devices need non-invertible Boolean memories, memories which can remember the past but not the future. So we arrive at the conclusion that the measurement problem can be solved only if

(1)　the basic theory can explain the breakdown of the endophysical time-inversion symmetry,

(2)　the basic theory can explain the existence of classical objects,

(3)　the basic theory can explain irreversible behavior,

(4)　the basic theory can explain the existence of nonanticipating memories,

(5)　the basic theory can explain the causality principle of the engineers saying that "response comes after stimulus".

Not one of these requirements can be realized in the framework of von Neumann's codification of quantum mechanics, but they can be fulfilled in more general codifications. According to algebraic quantum mechanics, these exophysical prerequisites may be *compatible* with the first principles of quantum endophysics, but they *certainly cannot be derived* from them. All so-called "solutions of the measurement problem" which are not explicitly introducing these or equivalent additional normative principles are wrong or at least not adequate.

For Bohr the quantum partition is a prerequisite for communication. If we accept the view that the only things we can talk about are *objects*, then it follows from algebraic quantum mechanics that an exosystem necessarily must have a classical description. Nevertheless, not every classical exosystem is appropriate for performing measurements. From a theoretical point of view, the challenge of the measurement problem for the theoreticians is to prove that the normative postulates of engineering exoscience are compatible with the first principles of endophysics. This is not a trivial affair. This problem is solved if and only if we can construct an appropriate GNS-representation of universally valid (but non-operational) quantum endophysics which satisfies the mentioned five normative conditions.

The dynamics of the exophysical measurement process has to be derived from the intrinsic endophysical dynamics. Often, measurements of the first kind are considered to have a duration so short that it can be taken as instantaneous. Obviously, such an idealization cannot be maintained in a discussion of the dynamical aspects of the measuring operation. A measuring-type process is due to the interaction of a quantum object with its classical environment (e.g. the classical measuring instrument) and has to be described *dynamically* on the basis of general first principles of quantum endophysics. Von Neumann's projection postulate implies the existence of a constant of motion which precludes a dynamical measuring process. However, a realization of almost-repeatable measurements in terms of probability operature measures is feasible.

Algebraic quantum mechanics in its individual interpretation implies that the intrinsic dynamics of an open quantum object maps pure states into pure states. If the object without its environment has only finitely many degrees of freedom, its pure state at time t can be represented in the usual way by a state vector $|\chi_t\rangle$, so that the reduced dynamics of the object is given by a map $|\chi\rangle \to |\chi_t\rangle$, $t > 0$. It can be shown quite generally that the corresponding equation of motion is a non-autonomous nonlinear functional–differential equation. In particularly simple situations this equation reduces to a non-linear stochastic Schrödinger–Langevin equation. In principle, such dynamical systems can describe the irreversible behavior of single objects, the so-called «quantum jumps» and the so-called «state reduction» in the measurement process.

5.2 An illustrative example

It may be helpful to illustrate the general considerations by an example for a measurement process. In a rather drastic caricature, we consider a spin-$\frac{1}{2}$, interacting linearly with a measurement instrument, overidealized as an infinite system of harmonic oscillators. The Hamiltonian

$$H = \tfrac{1}{2}\hbar\,\Omega\,\sigma\otimes 1 + 1\otimes \int_0^\infty \hbar\,\omega_k a_k^* a_k\,dk - \tfrac{1}{2}\hbar\,\sigma_3\otimes A \ , \tag{19}$$

with

$$A = \int_0^\infty \{\lambda_k^* a_k + \lambda_k a_k^*\}\,dk \ , \tag{20}$$

where $\omega_k > 0$, $\Omega = (\Omega_1, 0, \Omega_3) \in \mathbb{R}^3$, $\lambda_k \in \mathbb{C}$. The boson operators a_k fulfil the commutation relations

$$[a_k, a_{k'}] = 0 \ , \quad [a_k, a_k^*] = \delta(k-k') \ , \tag{21}$$

while the Pauli matrices σ_1, σ_2, σ_3 obey the commutation relations

$$[\sigma_\mu, \sigma_\nu] = 2i \sum_{\alpha=1}^3 \varepsilon_{\mu\nu\alpha}\sigma_\alpha \ . \tag{22}$$

The C*-algebra $\mathcal{A}$ of *intrinsic* observables is given by

$$\mathcal{A} = \mathcal{A}_s \otimes \mathcal{A}_{\text{meas}} \ , \tag{23}$$

where $\mathcal{A}_s$ is the C*-algebra of (2×2)-matrices, and $\mathcal{A}_{\text{meas}}$ is the C*-algebra of the intrinsic observables of the infinite measuring system. We choose $\mathcal{A}_{\text{meas}}$ to a C*-algebra of quasilocal observables, i.e. the norm closure of an inductive limit of subsystems with finitely many degrees of freedom. In our model, we assume that this C*-system represents the relevant endophysical system. The C*-algebra $\mathcal{A}_{meas}$ has a trivial center, so that in this endophysical system the quantum-mechanical superposition principle has unrestricted validity, and there are no intrinsic classical observables. In order to get an operational exophysical description of this system, we have to construct a faithful W*-representation. Since the C*-algebra $\mathcal{A}_{meas}$ is antiliminal (compare (Bratteli and Robinson, 1979, p.346)), there are infinitely many faithful representations and we have to select the one which represents the normative rules of present-day science and the abstractions made by the experimenter. In some (but not all) W*-representations of the C*-algebra $\mathcal{A}$ the Hamiltonian (19) induces a time evolution. As a rule, this dynamics cannot be implemented by a unitary group acting on the Hilbert space of the W*-representation. Together with the preconditions listed in sect. 5.1, we get the following conditions which have to be fulfilled in every physically acceptable W*-representation:

   (i)   The formal Hamiltonian (19) must induce a well-defined dynamics for
          every instant t > 0.
   (ii)   The dynamics of the W*-system has to be backward-deterministic and
          forward completely nondeterministic.
   (iii) The environment of the spin system must have a large classical part.
   (iv)   The environment of the spin system must possess a Boolean memory.
   (v)   The projection-value postulate has to be fulfilled within experimentally
          reasonable limits.

These conditions cannot be fulfilled for every choice of the parameters in the Hamitonian (19). The nature of the interaction between the object system (spin $\frac{1}{2}$) and its environment (the measurement instrument) is to a large extent determined by the following function $t \rightarrow K(t)$

$$K(t) := \int_0^\infty \frac{|\lambda_k|^2}{\omega_k} \cos(\omega_k t)\, dk \quad , \quad t \in \mathbb{R} \quad , \tag{24}$$

which depends only on the parameters occurring in the Hamiltonian and represents the relaxation properties of the combined system. The Hamiltonian dynamics is well defined if $K(0) < \infty$. In this case, the function $t \rightarrow K(t)/K(0)$ is the characteristic function of a probability distribution function. According to Lebesgue's decomposition theorem, there exists a unique decomposition of characteristic functions into a part coming from a purely discrete distribution, a part coming from an absolutely continuous distribution, and a part coming from a singular distribution. The dissipative and irreversible behavior, characterized by condition (ii) can be fulfilled only in the case of an absolutely continuous distribution, in this case the Riemann-Lebesgue lemma implies that

$$\lim_{t \to \infty} K(t) = 0 \quad . \tag{25}$$

The selection of a preferred time direction is effectuated by choosing the *retarded* solution of the differential equations of motion.

The simplest physically reasonable example (which can be justified as the first Padé-approximation to the memory kernel) is the Cauchy distribution, whose characteristic function is an exponential, so that

$$K(t) = \gamma \exp(-|t|/\tau) , \quad 0<\gamma<\infty , \quad 0<\tau<\infty . \tag{26}$$

In most practical situations the correlation time τ is very short in comparison to the other dynamical times, so that one may be interested only in the dynamics for $t \gg \tau$. Because of horrendous mathematical difficulties, we are not able to obtain a general idea of all W*-representations which fulfill these requirements. However, a special solution is governed for $t \gg \tau$ by the following simple system of three nonlinear stochastic differential equations in the sense of Stratonovich[1]

$$dM_1(t) = \{-\Omega_3 M_2(t) - \eta \Omega_1 M_2^2(t) + \gamma M_3(t)M_2(t)\}dt + \sqrt{2D}\, M_2(t)\circ dW(t) , \tag{27a}$$

$$dM_2(t) = \{\Omega_3 M_1(t) - \Omega_1 M_3(t) + \eta \Omega_1 M_1(t)M_2(t) - \gamma M_3(t)M_1(t)\}dt$$
$$- \sqrt{2D}\, M_1(t)\circ dW(t) , \tag{27b}$$

$$dM_3(t) = \Omega_1 M_2(t)dt . \tag{27c}$$

where γ, $\eta := \gamma\tau$, and D are three independent positive parameters. The symbol $\circ$ denotes the Stratonovich multiplication. Clearly, $M^2 = M(t)M(t)$ is a constant of motion, so that these equations in fact transform pure states into pure states. If we define a state vector $\psi_t \in \mathbb{C}^2$ by

$$M(t) = \langle \psi_t | \sigma | \psi_t \rangle , \tag{28}$$

and if we use the parametrization (14),

$$\psi_t = |\vartheta_t, \varphi_t\rangle , \tag{29}$$

we can describe the motion of the spin state vector as a trajectory $\{\vartheta(t), \varphi(t) \,|\, t \geq 0\}$ on the unit sphere in $\mathbb{R}^3$ by the following two stochastic differential equations:

$$d\vartheta_t = -\Omega_1 \sin(\varphi_t)\, dt , \tag{30a}$$

$$d\varphi_t = -\Omega_1 \cot(\vartheta_t)\cos(\varphi_t)\, dt + \Omega_3\, dt - \gamma \cos(\vartheta_t)\, dt$$
$$+ \eta\Omega_1 \sin(\vartheta_t)\sin(\varphi_t)\, dt - \sqrt{2D}\, dW(t) . \tag{30b}$$

The corresponding *nonlinear Schrödinger-Langevin equation*

$$d\psi_t = -\tfrac{1}{2}\{\mathbf{\Omega}\sigma + \eta\Omega M_2(t)\sigma_3 - \gamma M_3 \sigma_3\}\, \psi_t\, dt - \sqrt{2D}\, \sigma_3 \psi_t \circ dW(t) , \tag{31}$$

has a simple intuitive physical meaning. If we imagine that the elementary system with spin $\tfrac{1}{2}$ has the gyromagnetic ratio g, then we can write

$$i\hbar \frac{\partial \psi_t}{\partial t} = -\frac{g\hbar}{2} B(t)\sigma\, \psi_t , \tag{32}$$

---

[1]   For simple heuristic derivation of this equation, compare Primas (1990 b).

and interpret $B(t)$ as an effective magnetic field which acts at time t on the spin system. In the spirit of the Stratonovich calculus we assume that $dW(t)/dt$ exists in the ordinary sense as "almost white noise", so that we can write

$$gB(t) = \left(\Omega_1, 0, \Omega_3 + \eta \Omega_1 M_2 - \gamma M_3 - \sqrt{2D}\, \dot{W}(t) \right) \qquad (33)$$

In retrospect, this decomposition of the effective field $B(t)$ can be interpreted as follows:

(i) $g^{-1}\Omega$ is an external static magnetic field,

(ii) $g^{-1}\gamma M_3$ is an internal magnetic Onsager-type reaction field, due to the instantaneous polarization of the environment and an instantaneous feedback mechanism,

(iii) $g^{-1}\eta \Omega_1 M_2$ is a magnetic Onsager-type reaction field due to a dynamical polarization and a dynamical feedback mechanism (e.g. the Faraday induction law),

(iv) $g^{-1}\sqrt{2D}\, \dot{W}(t)$ is a classical fluctuating magnetic field, a direct influence of the environment on the object system.

For a proper choice of models of this kind (our simple example is restricted by the fact that there are only five disponible parameters, Ω_1, Ω_3, η, γ and D), such genuinely nonlinear stochastic Schrödinger equations can describe the main features of the quantum-mechanical measurement process. If the initial state vector for $t = 0$ is given by $|\vartheta_0, \varphi_0\rangle$, then the time evolution of the state vector can be described by the trajectory $(\vartheta_0, \varphi_0) \rightarrow (\vartheta_t, \varphi_t)$ on the unit sphere. The long-time behavior of this motion depends in an extremely sensitive way on the initial conditions of the environment. After a short initial time, the trajectory is attracted by either $\vartheta = 0$ ("spin up") or by $\vartheta = \pi$ ("spin down"). For a long time, the probability that the trajectory switches from a neighborhood of $\vartheta = 0$ to a neighborhood of $\vartheta = \pi$ can be arranged to be very small. In order to construct a good measuring instrument, the parameters of the Hamiltonian (19) have to be chosen in such a way that the measuring procedure is approximately reproducible, i.e. that the probability that after a short initial phase, the state is for a long time in the neighborhood of $\vartheta = 0$ becomes approximately $|c_\alpha|^2 = \cos^2(\vartheta/2)$. The smaller we choose the constant Ω_1, the nearer we can get to the result postulated by the projection postulate, but the whole measurement process also will become slower. In the limit $\Omega_1 \rightarrow 0$ the observable σ_3 becomes a constant of motion, and the measuring dynamics freezes.

This example is a drastically oversimplified[1] model of a measuring system, nevertheless it shows where the irreducible probabilities of quantum mechanics are coming from. The concept of probability does not appear in our starting point, quantum endophysics. *The probabilities enter the picture at the interface between the quantum object and the classical exosystem, they are contextual but inevitable and irreducible.* The deterministic motion $\psi \rightarrow \psi_t$ is chaotic. That is, even if the initial data of the object system

---

[1] An exosystem made of infinitely many harmonic oscillators can be used to elucidate the emergence of superselection rules, of classical properties, and of irreversibility, but it is too simple to model a permanent memory.

are perfectly known, the motion is unpredictable for times much longer than the correlation time of the perturbing exophysical classical *K-flow-type dynamical exosystem.*

To summarize: The measurement problem is basically the question how classical phenomena can be explained in terms of quantum physics, and how exophysical descriptions are related to first principles of endophysics. This problem is not of urgent interest in engineering science, but is important if we would like to know "how the world ticks". The Hilbert-space formalism of traditional quantum mechanics is too narrow to pose the problem in a proper way, yet within an appropriate mathematical codification of quantum theory this problem can be attacked even though it leads to formidable mathematical problems. Of course, "measurement-type" processes are not necessarily associated with quantal measurements but occur more typically also without any action induced by an experimentalist or observer.

REFERENCES

Ali, S. T. and G. G. Emch. (1974). *Fuzzy observables in quantum mechanics.* J. Math. Phys. **15,** 176–182.

Amrein, W. O. (1969). *Localizability for particles of mass zero.* Helv. Phys. Acta **42,** 149–190.

Arthurs, E. and J. L. Kelly. (1965). *On the simultaneous measurement of a pair of conjugate observables.* Bell System Tech. J. **44,** 725–729.

Benioff, P. A. (1972a). *Operator valued measures in quantum mechanics: finite and infinite processes.* J. Math. Phys. **13,** 231–242.

Benioff, P. A. (1972b). *Decision procedures in quantum mechanics.* J. Math. Phys. **13,** 908–915.

Benioff, P. A. (1972c). *Procedures in quantum mechanics without von Neumann's projection axiom.* J. Math. Phys. **13,** 1347–1353.

Benioff, P. (1973). *On definitions of validity applied to quantum theories.* Foundations of Physics **3,** 359–379.

Bratteli, O. and D. W. Robinson. (1979). *Operator Algebras and Quantum Statistical Mechanics. C*- and W*-Algebras, Symmetry Groups, Decomposition of States.* Vol. **1.** New York. Springer.

Davies, E. B. (1970). *On the repeated measurement of continuous observables in quantum mechanics.* J. Functional Analysis **6,** 318-346.

Davies, E. B. (1976). *Quantum Theory of Open Systems.* London. Academic Press.

Davies, E. B. and E. T. Lewis. (1970). *An operational approach to quantum probability.* Commun. Math. Phys. **17,** 239-260.

Finkelstein, D. (1988). *Finite physics.* In: *The Universal Turing Machine. A Half-Century Survey.* Ed. by R. Herken. Hamburg. Kammerer & Unverzagt. Pp. 349–376.

Gordon, J. P. and W. H. Louisell. (1966). *Simultaneous measurement of noncommuting observables.* In: *Physics of Quantum Electronics.* Ed. by P. L. Kelley, B. Lax and P. E. Tannenwald. New York. McGraw-Hill. Pp. 833–840.

Helstrom, C. W. (1976). *Quantum Detection and Estimation Theory.* New York. Academic Press.

Holevo, A. S. (1973a). *Statistical decision theory for quantum systems.* J. Multivariate Analysis **3,** 337–394.

Holevo, A. S. (1973b). *Statistical problems in quantum physics.* In: *Proceedings of the Second Japan-USSR Symposium on Probability Theory.* Ed. by G. Maruyama and Y. V. Prokhorov. Berlin. Springer. Pp. 104–119.

Holevo, A. S. (1974). *Optimal quantum measurements.* Theoretical and Mathematical Physics **17,** 1172–1177.

Holevo, A. S. (1978). *Investigations in the general theory of statistical decisions.* Proc. Steklov Institute Math. **Issue 3,** 1–140 (Russian number 124, 1976, pp.1–135).

Holevo, A. S. (1982). *Probabilistic and Statistical Aspects of Quantum Theory*. Amsterdam. North-Holland.

Jauch, J. M. and C. Piron. (1967). *Generalized localizability*. Helv. Phys. Acta **40**, 559–570.

Kraus, K. (1977). *Position observables of the photon*. In: *The Uncertainty Principle and Foundations of Quantum Mechanics*. Ed. by W. C. Price and S. S. Chissick. London. Wiley. Pp. 293–320.

Lüders, G. (1951). *Über die Zustandsänderung durch den Messprozess*. Annalen der Physik **8**, 322–328.

Ludwig, G. (1970). *Deutung des Begriffs "physikalische Theorie" und axiomatische Grundlegung der Hilbertraumstruktur der Quantenmechanik durch Hauptsätze des Messens*. Berlin. Springer.

Ludwig, G. (1983). *Foundations of Quantum Mechanics I*. New York. Springer.

Ludwig, G. (1985a). *An Axiomatic Basis for Quantum Mechanics. Volume 1. Derivation of Hilbert Space Structure*. Berlin. Springer.

Ludwig, G. (1985b). *Foundations of Quantum Mechanics II*. New York. Springer.

Ludwig, G. (1987). *An Axiomatic Basis for Quantum Mechanics. Volume 2. Quantum Mechanics and Macrosystems*. Berlin. Springer.

Neumann, J. v. (1927). *Wahrscheinlichkeitstheoretischer Aufbau der Quantenmechanik*. Nachr. Ges. Wiss. Göttingen, Math. Phys. Kl. **1927**, 245–272.

Neumann, J. v. (1932). *Mathematische Grundlagen der Quantenmechanik*. Berlin. Springer.

Ozawa, M. (1984). *Quantum measuring processes of continuous observables*. J. Math. Phys. **25**, 79–87.

Pauli, W. (1933). *Die allgemeinen Prinzipien der Wellenmechanik*. In: *Handbuch der Physik*. Ed. by H. Geiger and K. Scheel. Second ed.. Berlin. Springer.

Pauli, W. (1954). *Wahrscheinlichkeit und Physik*. Dialectica **8**, 112–124.

Peres, A. (1974). *Quantum measurements are reversible*. Amer. J. Phys. **42**, 886–891.

Primas, H. (1987). *Contextual quantum objects and their ontic interpretation*. In: *Symposium on the Foundations of Modern Physics, 1987. The Copenhagen Interpretation 60 Years after the Como Lecture*. Ed. by P. Lahti and P. Mittelstaedt. Singapore. World Scientific. Pp. 251–275.

Primas, H. (In press). *Time-asymmetric phenomena in biology. Complementary exophysical descriptions arising from deterministic quantum endophysics*. In: *Proceedings of the International Workshop "Information Biothermodynamics"*. Torun.

Primas, H. (1990a). *Mathematical and philosophical questions in the theory of open and macroscopic quantum systems*. In: *Sixty-two Years of Uncertainty: Historical, Philosophical and Physics Inquiries into the Foundations of Quantum Mechanics*. Ed. by A. I. Miller. New York. Plenum.

Primas, H. (1990b). *Induced nonlinear time evolution of open quantum objects*. In: *Sixty-two Years of Uncertainty: Historical, Philosophical and Physics Inquiries into the Foundations of Quantum Mechanics*. Ed. by A. I. Miller. New York. Plenum.

Raggio, G. A. (1981). *States and Composite Systems in W*-algebraic Quantum Mechanics*. Thesis ETH Zürich, No. 6824, ADAG Administration & Druck AG, Zürich.

Rényi, A. (1966). *On the amount of missing information and the Neyman-Pearson lemma*. In: *Research Papers in Statistics. Festschrift for J. Neyman*. Ed. by F. N. David. London. Wiley. Pp. 281–288.

Rössler, O. E. (1987). *Endophysics*. In: *Real Brains, Artificial Minds*. Ed. by J. L. Casti and A. Karlqvist. New York. North-Holland. Pp. 25–46.

She, C. Y. and H. Heffner. (1966). *Simultaneous measurement of noncommuting observables*. Phys. Rev. **152**, 1103–1110.

Shimony, A. (1979). *Proposed neutron interferometer of some nonlinear variants of wave mechanics*. Phys. Rev. A **20**, 394–396.

Shimony, A. (1984). *The Quantum World*. Manuscript for a yet unpublished book, distributed at a seminar at the ETH Zürich, Nov. 1984.

Srinivas, M. D. (1980). *Collapse postulate for observables with continuous spectra*. Commun. Math. Phys. **71**, 131–158.

Weinberg, S. (1989). *Testing Quantum Mechanics*. Annals of Physics **194**, 336–386.

Principle of Stationarity in the Action Functional and the Theory of Measurement

R. FUKUDA

1. FLUCTUATION AND MACROVARIABLES

The measurement problem in quantum mechanics is certainly one of the profound problems in the theoretical physics. It has a long history and for the present status we refer for instance to the proceedings of the 2nd ISQM edited by Namiki et.al. (1986). Those of the 3rd ISQM will appear soon (1989). In this paper we start from the following observation which is somewhat different from the conventional approach.

The classical Newtonian mechanics does not have any measuring problem, any dynamical variables $q_i(i=1\sim N)$ satisfying the deterministic equation of motion. If the system is discussed quantum mechanically, q_i satisfies the same equation of motion as the classical one but it is now the operator equation since q_i is an operator due to the canonical commutation relation between q_i and the momentum p_i. The non-commutativity is the source of the quantum fluctuations and these uncontrollable fluctuations are the cause of the measurement problem in quantum mechanics.

If we can find the fluctuationless variables in quantum system, the problem of measurement becomes parallel with that of the classical system. Therefore the solution to our measurement problem is *the search for the fluctuationless variables* in quantum system.

Along this line the measurement theory has been presented by the author (1987a, b, 1988, 1989a, b). We can see, in this formalism, the reason why the detector is usually the macroscopic system: the macroscopic system has in itself fluctuationless variables naturally. We call these variables macrovariables which are defined as an average over the macroscopic number of microscopic variables.

In the following macrovaribles are denoted by X symbolically. The center of mass co-ordinate is the simplest example: $X = \frac{1}{N}\sum_{i=1}^{N} q_i$. It is well known that X loses fluctuations as $N \to \infty$.

Let us compare the classical limit $\hbar \to 0$ and the macroscopic limit $N \to \infty$. The following path integral formula relates the two wave functions $\psi_a(q_i)$ and $\psi_b(q_i)$ at the time $t = t^a$ and t^b respectively;

$$\psi_b(q_i) = \int_B [dq] \exp\frac{i}{\hbar} \int dt L(\dot{q}_i, q_i) \psi_a(q_i). \tag{1}$$

Here $\int_B [dq]$ is the path integral with the appropriate boundary conditions at $t = t^a$ and t^b and L is the Lagrangian of the system. The integration over q signifies that q_i is the fluctuating variable. Since $1/\hbar$ multiplies the whole action $\Gamma \equiv \int dt L$, in the limit $\hbar \to 0$ all the variables q_i trace the deterministic trajectory given by the stationary phase of the action functional. This is nothing but the classical equation of motion. The fluctuations are suppressed as $\sqrt{\hbar}$.

For the macroscopic system, let us separate q_i into the macrovariable X and the microvariables q_i' , the relative co-ordinates for example. In general X and q_i' do not separate and we have three terms in the action,

$$\Gamma = NI^{(1)}(X) + I^{(2)}(X, q_i') + I^{(3)}(q_i'). \tag{2}$$

The first term involves X only and has a factor N in front since the action is the extensive quantity of the order N . $I^{(2)}$ and $I^{(3)}$ involve the summation over i and $I^{(2)}$ is the cause of the dissipation of the variable X . By rewriting $[dq]$ in terms of the integal over X and q_i' , we see that only X loses fluctuations as $N \to \infty$ and takes the c-number value X_c . The microvariables q_i' remains fluctuating and construct the Hilbert space specified by the value X_c .

Now we use the macrovariable X as the detector variable, the pointer position for example. The fluctuating microscopic variables of the object is mapped onto the deterministic variable X in the course of the measurement. But this process requires *amplification*. The amplification mechanism is again studied by the stationarity condition of the action functional. We summarize our assertions in the following statement:

> Including amplification mechanism, the measurement problem is clarified by the principle of the stationarity in the action functional.

The interference term vanishes between the two states with different values of the c-number macrovariables since they belong to the different Hilbert space. This can be seen in our path integral method by the appearance of the peak at $X=X_c$ in the wave function which is proportional to the δ-function $\delta(X-X_c)$.

Two examples are given below illustrating the above ideas. The first one is described by the quantum mechanics of N-particle system and in the second example the field theoretical model is adopted for the detector.

2. MODEL 1 - THE CENTER OF MASS

When one studies the macroscopic system, it is easily analyzed if the relevant macrovariable is separable from the microvariables. The simplest example is the center of mass co-ordinate X . We take the one dimensional system and discuss below only the wave function of X .

We start from finite N. Suppose, at the time $t=t^a$, the wave function is given by $\psi^a(X)$ then at $t=t^b$ we find

$$\psi_b(X^b) = \int K(X^b,t^b;X^a,t^a)\psi_a(X^a)dX^a, \tag{3}$$

where K is given by the action functional $\Gamma[X]$ as

$$K = \int_B[dX]e^{i\Gamma[X]}, \quad \Gamma[X] = \int_{t^a}^{t^b}L(X)dt, \quad L(X) = \frac{Nm}{2}\dot{X}^2. \tag{4}$$

The path integral $\int_B[dX]$ has to satisfy the boundary conditions $X(t^a)=X^a$, $X(t^b)=X^b$ and it is assumed to be defined by including the appropriate normalization factor. Now we know that K is given by the diffusion kernel,

$$K = \sqrt{\frac{Nm}{2\pi i T}} e^{iS}, \tag{5}$$

$$S = \frac{Nm}{2T}(X^b - X^a)^2, \qquad T = t^b - t^a \tag{6}$$

In the limit $N \to \infty$, the stationary phase of e^{iS} dominates and we get $K = \delta(X^b - X^b)$, $\psi_b(X^b) = \psi_a(X^b)$, so that the wave function does not change. For finite N there appears the variance of the order $1/\sqrt{N}$. In case $\psi_a(X^a)$ has the form

$$e^{iP^a X^a} \psi(X^a), \tag{7}$$

with P^a a quantity of the order N, then the stationarity condition is changed and, for $N \to \infty$, $\psi_b(X^b)$ is found to be

$$\psi_b(X^b) = \exp(i\frac{P^{a2}}{2Nm}T)\int \delta(X^b - X^a - \frac{P^a}{Nm}T)\psi_a(X^a)dX^a \tag{8}$$

$$= \exp(i\frac{P^{a2}}{2Nm}T)\psi_a(X^b - \frac{P^a}{Nm}T). \tag{9}$$

Equations (8) and (9) imply the absence of the quantum diffusion and show that X^b is determined once X^a and T is fixed; X is the deterministic variable. We illustrate (9) in Fig.1 schematically. We see one to one correspondence between X^b and X^a showing the c-number character of the center of mass variable.

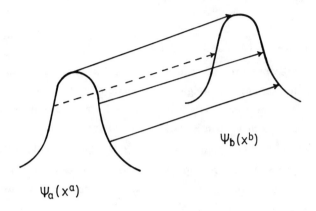

$\Psi_b(x^b)$

$\Psi_a(x^a)$

Fig.1 Behavior of the wave function of (9)

The equation of motion is obtained by the stationary phase of $e^{i\Gamma}$

$$0 = \frac{1}{N}\frac{\delta\Gamma\ [X]}{\delta X(t)} = -m\ddot{X}. \tag{10}$$

We need the boundary conditions at $t=t^b$ to solve (10). The velocity v^a at $t=t^a$ is, by (6),

$$v^a = \frac{-1}{Nm}\frac{\partial S}{\partial X^a} = \frac{X^b-X^a}{T} = \frac{P^a}{Nm} \equiv \frac{p}{m}. \tag{11}$$

With this and $X=X^a$ at $t=t^a$, we can solve (10) as $X^b = X^a + \frac{p}{m}T$. The parameter (X,p) that specifies the Hilbert space spanned by the relative coordinates is
given at the arbitrary time $t=t^b$ by $(X^b, p=\frac{P^a}{N})$, where p is time independent. We
can clearly see the fact that the points with different X^b does not interfere with
each other by looking at $|\psi_b(X^b)|^2$:

$$|\psi_b(X^b)|^2 = \int dX_1^a \int dX_2^a \delta(X^b - X_1^a - \frac{p}{m}T)\delta(X^b - X_2^a - \frac{p}{m}T)\psi_a(X_1^a)^*\psi_a(X_2^a)$$

$$= \int dX^a \delta(X^b - X^a - \frac{p}{m}T)|\psi_a(X^a)|^2.$$

The cross term between $\psi_a(X_1^a)$ and $\psi_a(X_2^a)$ vanishes if $X_1^a \neq X_2^a$.

Summarizing the above results, the state (7) is the *mixed state* and the functional form of X determines the weight with which the state lies in the Hilbert space specified by X. The weight is given of course by the absolute square of (7). If it is further superposed in p, then it is also a mixed state in p.

Now in order to shift the c-number value of X, the state like (7) has to be prepared and it requires the macroscopic energy. In some cases we get this energy by the object we are going to measure. Otherwise the amplification mechanism is necessary. This is the subject of the next section.

3. AMPLIFICATION BY EXTERNAL FIELD

In order to change X in time, let us discuss the mechanism to supply energy by the external field. We take as a model Lagrangian

$$L + L^I = \frac{Nm}{2}\dot{X}^2 + g\alpha X \tag{12}$$

where g is the coupling strength and α is the external field which is assumed to be time independent for simplicity. If it is of the macroscopic order, we rescale α as, $\alpha = N\hat{\alpha}$. Here α is, for example, the electric field if X has the charge. The kernel corresponding to (12) is given by (5) but with the following S (Feynman and Hibbs (1948)).

$$S = \frac{Nm}{2T}(X^b - X^a)^2 + \frac{g\alpha}{2}T(X^b + X^a) - \frac{g^2\alpha}{24mN}T^3. \tag{13}$$

Thus we have

$$\psi_b(X^b) = e^{i\theta}\sqrt{\frac{Nm}{2\pi iT}}\int dX^a \exp\{i\frac{Nm}{2T}(X^b - X^a - \frac{g\alpha}{2mN}T^2)^2\}\psi_a(X^a) \tag{14}$$

$$\underset{N\to\infty}{\longrightarrow} \; e^{i\theta}\int dX^a \delta(X^b - X^a - \frac{g\alpha}{2mN}T^2)\psi_a(X^a),\tag{15}$$

$$\theta = g\alpha T(X^b - \frac{g\alpha}{6mN}T^2).\tag{16}$$

The term $g\alpha TX^b$ in θ corresponds to the fact that the momentum increases by the amount $g\alpha T$ which means $p = g\alpha T/N$.

In terms of the action functional Γ, given by

$$\Gamma\,[X,\alpha] = \int_{t^a}^{t^b} dt(\frac{Nm}{2}\dot{X}^2 + g\alpha X),\tag{17}$$

we solve the deterministic equation

$$\frac{1}{N}\frac{\delta\Gamma}{\delta X(t)} = -m\ddot{X} + \frac{g\alpha}{N} = 0\tag{18}$$

with the boundary condition $X(t^a) = X^a$, $\dot{X}(t^a) = 0$. We get the same result of course; $X^b = X^a + \frac{g\alpha}{2mN}T^2$.

We stress here that the important factor in the measurement theory is not the infinite phase θ but the presence of the δ-function which implies the absence of the fluctuation for the macrovariable.

4. MEASUREMENT OF THE MICROSCOPIC OBJECT

The above system is now used as a detector. Although it is far from realistic we can see the essential points clearly.

Before measurement, at $t=t_0$ the wave function of the object or the detector is written as $\Phi_0(r) = \Phi_0^{(1)}(r) + \Phi_0^{(2)}(r)$, or ψ_0 respectively. Here we consider in three dimensions. We assume that at $t=t^a$, the object and the detector interact microscopically so that the macrovariable does not change. At the time $t=t^b$ the measurement has been performed so that there occurs a measurable shift in the macrovariable. We have therefore

Quantum Theory without Reduction

$$\Phi_0(\mathbf{r})\psi_0 = (\Phi_0^{(1)}(\mathbf{r}) + \Phi_0^{(2)}(\mathbf{r}))\psi_0 \tag{19}$$

$$\underset{t=t^a}{\rightarrow} \Phi_a^{(1)}(\mathbf{r})\psi_a^{(1)} + \Phi_a^{(2)}(\mathbf{r})\psi_a^{(2)} \tag{20}$$

$$\underset{t=t^b}{\rightarrow} \Phi_b^{(1)}(\mathbf{r})\psi_b^{(1)} + \Phi_b^{(2)}(\mathbf{r})\psi_b^{(2)}. \tag{21}$$

The factorized form of the wave function is assumed above. We are considering such an idealized experiment where we can neglect the influence of the microscopic interaction on the object wave function. A typical case is the one where the object passes near the detector in a well-defined form of the wave packet. In this case the factorized form will be a good approximation. We amplify the microscopic change in the detector wave function ψ_a.

Consider the Stern-Gerlach type experiment shown in Fig.2. The wave packet represented by the linear combination $\Phi_0^{(1)}(\mathbf{r}) + \Phi_0^{(2)}(\mathbf{r})$, each representing the up (down) state of the z-component of the spin, is separated by the magnetic field which has the large gradient in the z-direction. Our detector is assumed to be placed near the path of the packet 1.

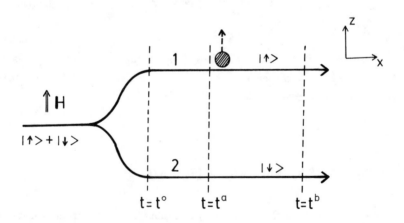

Fig.2 Experiment of the Stern-Gerlach type

The exact treatment of the many particle system is impossible so we proceed as follows. First the interaction between the object and the detector is assumed to be of the following type. The object is not affected by the external field, i.e. $g=0$. The detector also consists of the particles with $g=0$. They are changed, with some probability, into the charged state with $g \neq 0$ by the interaction with the object.

This is the microscopic, quantum mechanical coherent process. Let x_1 be the coordinate of this charged particle then its interaction with the external field is given by $g\alpha x_{1z}$, where the external field is taken along the z-direction. In terms of the center of mass coordinate X of the all particles, including neutral particles, this term is equivalent to $g\alpha X_z$. So that our total Lagrangian is

$$L = \frac{Nm}{2}\dot{X}^2 + g\alpha X_z + L_{rel.} \,, \tag{22}$$

where $L_{rel.}$ depends only on the relative coordinates. For the state of n charged particles we have

$$L = \frac{Nm}{2}\dot{X}^2 + ng\alpha X_z + L_{rel.} \,. \tag{23}$$

Now we discuss the wave function of the detector at $t=t^a$, which should really be written by $\psi_a(\xi_i)$. Here ξ_i represents the coordinate of every particle in the detector so that $\xi_i = (X,\zeta_i)$, where ζ_i is the relative coordinate. We do not have to distinguish between the charged or uncharged co-ordinates in the following discussions. Let ψ_n be the normalized wave function of n-charged sector and write for the path 1,

$$\psi_a^{(1)}(\xi_i) = \sum_{n=0} C_n\psi_n. \tag{24}$$

Since we are studying the microprocess here, the wave function of the center of mass is not affected by this process and at $t=t^a$ the common wave function $\psi_a(X)$ is assumed for both paths 1 and 2. So we can write, by the factorization of the center of mass,

$$\begin{aligned}
\psi_a^{(1)}(\xi_i) &= \sum_{n=0} C_n\widetilde{\psi}_n(\zeta_i)\psi_a(X), \quad \cdots \quad \text{for path 1,} \\
\psi_a^{(2)}(\xi_i) &= \widetilde{\psi}_0(\zeta_i)\psi_a(X), \qquad \cdots \quad \text{for path 2.}
\end{aligned} \tag{25}$$

Now after the elapse of the time, we get at $t=t^b$,

$$\psi_b^{(1)}(X^b,\zeta_i^b) = \sum_{n=0} C_n e^{i\theta_n} \int d^3 X^a \delta^2(X_\perp^b - X_\perp^a)\delta(X_z^b - X_z^a - \frac{ng\alpha}{2Nm}T^2)\psi_a(X^a)\widetilde{\psi}_{nb}(\zeta_i^b), \quad (26)$$

$$\theta_n = n\alpha gT(X^b - \frac{ng\alpha}{6mN}T^2), \quad (27)$$

$$\psi_b^{(2)}(X^b,\zeta_i^b) = \int d^3 X^a \delta^3(X^b - X^a)\psi_a(X^a)\widetilde{\psi}_{0b}(\zeta_i^b), \quad (28)$$

where we have introduced $X_\perp = (X_x, X_y)$.

Integrating over ζ_i^b after the insertion of the arbitrary function $P(\zeta_i^b)$, we find,

$$I \equiv \int d\zeta_i^b P(\zeta_i^b)|\phi_b^{(1)}(r)\psi_b^{(1)}(X^b,\zeta_i^b) + \phi_b^{(2)}(r)\psi_b^{(2)}(X^b,\zeta_i^b)|^2 \quad (29)$$

$$= A_{11}|\phi_b^{(1)}(r)|^2 + A_{22}|\phi_b^{(2)}(r)|^2$$

$$+ \{A_{12}\phi_b^{(2)*}(r)\phi_b^{(1)}(r) + \text{complex conj.}\}, \quad (30)$$

$$A_{11} = \sum_{l,n=0} D_{l,n} C_n^* C_l e^{i(\theta_l - \theta_n)}$$

$$\times \int dX_z^{a'} dX_z^a \delta(X_z^b - X_z^{a'} - \frac{ng\alpha}{2Nm}T^2)\delta(X_z^b - X_z^a - \frac{lg\alpha}{2Nm}T^2) \quad (31)$$

$$\times \psi_a^*(X_\perp^b, X_z^{a'})\psi_a(X_\perp^b, X_z^a),$$

$$A_{22} = D_{00}|\psi_a(X^b)|^2, \quad (32)$$

$$A_{12} = D_{00}C_0 A_{22} + \sum_{n\neq 0} D_{n0} C_n e^{i\theta_n} \int dX_\perp^a dX_z^{a'} dX_z^a \delta(X_\perp^b - X_\perp^a)\delta(X_z^b - X_z^{a'} - \frac{ng\alpha}{2Nm}T^2)$$

$$\times \psi_a^*(X_\perp^a, X_z^a)\psi_a(X_\perp^{a'}, X_z^{a'}), \quad (33)$$

$$D_{l,n} = \int d\zeta_i P(\zeta_i^b)\widetilde{\psi}_n^*(\zeta_i)\widetilde{\psi}_l(\zeta_i). \quad (34)$$

Note here that the presence of arbitrary polynomial $P(\zeta_i^b)$ implies that the structures of (31), (32) and (33) hold without the integration over ζ_i; they are not related to the orthogonality relations of $\widetilde{\psi}_n(\zeta_i)$, in particular not to the superselection rules (due to the charge conservation law for example).

Now let us study what controls the magnitude of the interference term A_{12}.

1) C_0; This factor naturally comes in and it remains even when the mechanisms about the macrovariables of the detector are ideally arranged. In order to enhance the detection sensitivity, we must make C_0 small by choosing, for instance, the material which is easily 'ionized' into the charged states.

2) The overlap of the states corresponding to different values of the macrovariables: Let us assume $\psi_a = \Psi_a(X_\perp)\Psi_a(X_z)$ and take

$$\Psi_a(X_z) = \sqrt{\frac{1}{\sqrt{\pi}\,a}}\exp(-\frac{X_z^2}{2a^2}). \tag{35}$$

We further replace $\delta(X_z^b - X_z^a)$ in the second term of (33) by its expression of finite N,

$$\sqrt{\frac{Nm}{2\pi i T}}\exp\frac{iNm}{2T}(X_z^b - X_z^a)^2 \tag{36}$$

and perform the integration. Then the absolute value of one of the terms in the sum appearing in (33) is proportional to

$$\frac{1}{1 + \dfrac{T^2}{a^4 N^2 m^2}}\sqrt{\frac{1}{\sqrt{\pi}\,a}}\exp\{-\frac{X^{b^2} + (X_z^b - \Delta X_z)^2}{2a^2(1+\rho)}\}, \tag{37}$$

$$\Delta X_z = \frac{gn\alpha}{2Nm}T^2, \quad \rho = (\frac{T}{aNM})^2. \tag{38}$$

Equation (37) becomes maximum at $X_z^b = \dfrac{\Delta X_z}{2}$ with the value

$$\frac{1}{1+\dfrac{T^2}{a^4 N^2 m^2}} \sqrt{\frac{1}{\sqrt{\pi} a}} \exp\{-\frac{\Delta X_z^2}{4a^2(1+\rho)}\}. \tag{39}$$

In the limit $N \to \infty$, magnitude of $\Delta X_z / a$ determines the interference term (39) in the exponential manner $\exp(-\dfrac{\Delta X_z^2}{4a^2})$. The ideal limit is $N \to \infty$, $\Delta X_z / a \to \infty$. Let us see what makes ΔX_z large. We need the large $(gn)(\dfrac{\alpha}{N})T^2$. Here the factor gn is determined by the property of the detector and the large $\dfrac{\alpha}{N} = \hat{\alpha}$ is realized by applying macroscopically large external field. The large T is also preferable but N should be large in the first place and then T can be taken to be large. Summarizing, the ideal limit to get the small interference is

$$N \to \infty, \quad C_0 \to 0, \quad \Delta X_z / a \to \infty, \quad (\text{then} \quad T \to \infty). \tag{40}$$

We can easily see that the same limit is the limit of the maximum sensitivity; clear shift of X_z leads to the disappearance of the interferences. Similar relation has been discussed with different models by Cini (1982).

We need non-vanishing ΔX_z anyhow in the measuring process. If α is of the order unity, T^2 should be of the order N: the infinite time is required for the measurement (Hepp, (1972)). *Amplifications realized by taking $\alpha = O(N)$ make it possible to perform the measurement in the finite time interval.*

5. MODEL 2 - GENERALIZED MACROVARIABLES

The macrovariables are defined up to now to involve an average over the macroscopic number of microvariables. But since our essential property of the macrovariables are the loss of fluctuations, any variable can play the role once the stationary phase appears in the action functional when we integrate over these variables. An example is given by taking a field theoretical model for the detector.

The object particle with the co-ordinate r passes through the detector described by the second quantized field variable $\phi(x)$. The total Lagrangian of the model is given by

$$L = \int dx \{ \frac{m}{2} \dot{\phi}(x)^2 - \frac{m\omega^2}{2} \phi(x)^2 - \frac{m\kappa}{2} (\nabla \phi(x))^2 + g\phi(x)\rho(x-r(t)) \}. \qquad (41)$$

Here $\rho(x)$ is the function representing the range of the interaction between the object and the field. We assume that $\rho(x)$ has the support in the small region of microscopic size near $x=0$. The dispersion $\omega_k{}'$ of the field is assumed for simplicity to be $\omega_k{}'^2 = \omega^2 + \kappa k^2$.

The physical picture of the model is the following. The particle proceeds in the detector shifting the value of $\phi(x)$ around it losing its energy. We apply the external field α to supply the energy to the particle. In the limit $m \to 0$, for example, the shift of $\phi(x)$ becomes infinite and we get the generalized macrovariable $m\phi(x)$. The limit $m \to 0$ makes the ϕ-system to be easily excited (easily kicked) by the interaction with the particle. The strong external field has thus to be supplied for small m. So our idealized limit is $m \to 0$, $\alpha \to \infty$.

The model (41) cannot exactly be solved therefore we first solve $\phi(x)$ by the fixed $r(t)$ and then $r(t)$ is solved by using the solution of $\phi(x)$. Equation (41) is written in Fourier components as

$$L = \sum_{k,\sigma}{}' \{ m\dot{\phi}_k^{\sigma 2} - m\omega_k{}'^2 \phi_k^{\sigma 2} \}$$

$$+ 2g\sum_k{}' \{ \mathrm{Re}(\phi_k \rho_{-k}) \cos k \cdot r(t) - \mathrm{Im}(\phi_k \rho_{-k}) \sin k \cdot r(t) \} . \qquad (42)$$

Here $\sigma = R,I$ and $\mathrm{Re}(Im)$ means the real (imaginary) part. We take $\rho(x)$ to be a function of x^2, so that $\rho_k = \rho_{-k} \equiv \rho_k$ which is real. The symbol $\sum'$ signifies the summation over the half k-space because of the symmetry property of ϕ_k^σ .

The detector is assumed to be in the ground state at $t=t^a$. The wave functional is given by

$$\psi_a[\phi_k^a] = \prod_{k,\sigma}{}' (\frac{2m\omega_k{}'}{\pi})^{\frac{1}{4}} \exp(-m\omega_k{}' \phi_k^{\sigma a2}) . \qquad (43)$$

At $t=t^b$ we have

$$\psi_b[\phi_{\boldsymbol{k}}^b] = \int K[\phi_{\boldsymbol{k}}^b, t^b; \phi_{\boldsymbol{k}}^a, t^a; r] \psi_a[\phi_{\boldsymbol{k}}^a] \prod_{\boldsymbol{k},\sigma}{}' d\phi_{\boldsymbol{k}}^{\sigma a} . \tag{44}$$

By L of (42), K can be calculated to be (Feynman and Hibbs (1948), Fukuda (1987b)), with $T = t^b - t^a$,

$$K = \prod_{\boldsymbol{k}}{}' (\frac{2m\omega_{\boldsymbol{k}}{}'}{\pi})^{\frac{1}{2}} (\frac{m\omega_{\boldsymbol{k}}{}'}{i\pi\sin\omega_{\boldsymbol{k}}{}' T}) \exp\{i\sum{}' \frac{m\omega_{\boldsymbol{k}}{}'}{\sin\omega_{\boldsymbol{k}}{}' T}(S_{\boldsymbol{k}}^R + S_{\boldsymbol{k}}^I)\} , \tag{45}$$

$$\begin{aligned}
S_{\boldsymbol{k}}^R = {}& \cos\omega_{\boldsymbol{k}}{}' T(\phi_{\boldsymbol{k}}^{Ra2} + \phi_{\boldsymbol{k}}^{Rb2}) - 2(\sin\omega_{\boldsymbol{k}}{}' T)\phi_{\boldsymbol{k}}^{Ra}\phi_{\boldsymbol{k}}^{Rb} \\
& + \frac{2g\rho_{\boldsymbol{k}}}{m\omega_{\boldsymbol{k}}{}'}\phi_{\boldsymbol{k}}^{Rb}\int_{t^a}^{t^b} dt\ \cos\boldsymbol{k}{\cdot}r(t)\sin\omega_{\boldsymbol{k}}{}'(t-t^a) \\
& + \frac{2g\rho_{\boldsymbol{k}}}{m\omega_{\boldsymbol{k}}{}'}\phi_{\boldsymbol{k}}^{Ra}\int_{t^a}^{t^b} dt\cos\boldsymbol{k}{\cdot}r(t)\sin\omega_{\boldsymbol{k}}{}'(t^b-t) \\
& - \frac{2g^2\rho_{\boldsymbol{k}}^2}{m\omega_{\boldsymbol{k}}{}'}\int_{t^a}^{t^b} dt\int_{t^a}^{t} ds\cos\boldsymbol{k}{\cdot}r(t)\cos\boldsymbol{k}{\cdot}r(s)\sin\omega_{\boldsymbol{k}}{}'(t^b-t)\sin\omega_{\boldsymbol{k}}{}'(s-t^a) ,
\end{aligned} \tag{46}$$

$$S_{\boldsymbol{k}}^I = S_{\boldsymbol{k}}^R(\cos\boldsymbol{k}{\cdot}r(t) \rightarrow -\sin\boldsymbol{k}{\cdot}r(t)). \tag{47}$$

Equations (43) and (45) are inserted into (44) and we perform the $\phi_{\boldsymbol{k}}^a$ integration. This can be done but what we are interested in is the dependence on $\phi_{\boldsymbol{k}}^b$. Apart from the phase factor it is given by

$$\begin{aligned}
\psi_b(\phi_{\boldsymbol{k}}^b) \propto {}& \prod_{\boldsymbol{k}}{}' \exp\{-m\omega_{\boldsymbol{k}}{}'(\phi_{\boldsymbol{k}}^{Rb}-\beta_{\boldsymbol{k}}^R)^2\} \\
& \times \prod_{\boldsymbol{k}}{}' \exp\{-m\omega_{\boldsymbol{k}}{}'(\phi_{\boldsymbol{k}}^{Ib}-\beta_{\boldsymbol{k}}^I)^2\} ,
\end{aligned} \tag{48}$$

$$\beta_{\boldsymbol{k}}^R = \frac{g\rho_{\boldsymbol{k}}}{m\omega_{\boldsymbol{k}}{}'}\int_{t^a}^{t^b} dt\cos\boldsymbol{k}{\cdot}r(t)\sin\omega_{\boldsymbol{k}}{}'(t^b-t) , \tag{49}$$

$$\beta_k^I = \frac{g\rho_k}{m\omega_k{}'}\int\limits_{t^a}^{t^b} dt \sin k \cdot r(t)\sin\omega_k{}'(t^b-t) \ . \tag{50}$$

Now consider the limits which bring us the stationary phase.

i)   By putting $\phi_k^{Rb}/g = \widetilde{\phi}_k^{Rb}$, one of the factors in (48) is written as

$$\exp\{-m\omega_k{}' g^2(\widetilde{\phi}_k^{Rb} - \frac{\beta_k^R}{g})^2\} \ . \tag{51}$$

The same holds for ϕ_k^I. $\widetilde{\phi}_k^{Rb}$ loses fluctuation around the finite value β_k^R/g as $g\to\infty$.

ii)   Writing in (48) $m\phi_k^{Rb} = \hat{\phi}_k^{Rb}$, we get the factor

$$\exp\{-\frac{\omega_k{}'}{m}(\hat{\phi}_k^{Rb} - m\beta_k^R)^2\} \ . \tag{52}$$

The same for ϕ_k^I. In the limit $m\to 0$, $\hat{\phi}_k^{Rb}$ becomes deterministic taking the finite value $m\beta_k^R$.

We consider above two cases in the following. In x-representation, $\frac{m}{g}\phi(x)$ tends to a c-number value $\gamma(x)$ for all x, where

$$\gamma(x) = \frac{2}{\sqrt{V}}\sum_k{}' \int\limits_{t^a}^{t^b} dt \cos k \cdot (x-r(t))\frac{\sin\omega_k{}'(t^b-t)}{\omega_k{}'}\rho_k \tag{53}$$

$$= \frac{1}{2\pi}\kappa^{-\frac{3}{2}}\int dx' \rho(x')\int\limits_{t^a}^{t^b} dt\{\delta(\xi^2) - \frac{\omega^2}{2}\theta(\xi^2)\frac{J_1(\omega\sqrt{\xi^2})}{\sqrt{\xi^2}}\} \ , \tag{54}$$

$$\xi^2 = (t^b-t)^2 - \kappa^{-1}(x-r(t)-x')^2 \ . \tag{55}$$

Here J_1 is the Bessel function. In Fig.3, the region is shown where $\frac{m}{g}\phi(x)$ is non-vanishing. We have assumed in the figure

$$r(t) = r(t^b) - (t^b-t)v \tag{56}$$

with some constant v. Around the axis of the trajectory of the particle, we see a shaded cone-like region satisfying $\xi^2 > 0$ where $\dfrac{m}{g}\phi(x)$ is a finite c-number. Since each k-mode is independent, there is no dissipation and $\gamma(x)$ given in (54) is oscillating as a function of t^b.

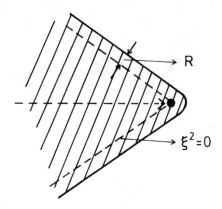

Fig.3 Experiment of the model 2. The shaded area is the region where $\gamma(x)$ of (53) is non-zero. The object particle is represented by a black disc. R denotes the region due to the spread of $\rho(x)$

Let us define $h(x)$ by

$$h(x) = \frac{m}{g}\phi(x) - \gamma(x)$$

then the right handside of (48) is written as

$$\exp\{-\frac{1}{2}\int dx \int dx'\, h(x)D(x-x')h(x')\} ,\tag{57}$$

where D^{-1} determines the fluctuation. By using $r=|x-x'|$, it is given by

$$D^{-1}(x-x') = \frac{1}{\pi}\kappa^{-\frac{3}{2}}\frac{m}{g^2}\frac{\sqrt{\kappa\omega}}{r}K_1(\frac{\omega}{\sqrt{\kappa}}r) \underset{r\to\infty}{\sim} e^{-\frac{\omega}{\sqrt{\kappa}}r} .\tag{58}$$

It is found that the fluctuation is of the order $\sqrt{\dfrac{m}{g^2}}$ and its correlation vanishes exponentially. (The limit $r\to 0$ is related to $k^2\to\infty$ and if we cut off the k-space at $k^2=\Lambda^2$, (58) becomes finite for $r\to 0$).

The next task is to consider the back reaction on the particle from the ϕ-field. We make approximations as follows. Consider the momentum p_b of the particle at $t=t_0$. Denoting the total Hamiltonian by H, we calculate

$$\dot{p}_b = i[H, p_b] \tag{59}$$

$$= -g\int d\boldsymbol{x}\phi(\boldsymbol{x})\frac{\partial}{\partial\boldsymbol{x}}\rho(\boldsymbol{x}-\boldsymbol{r}(t^b)) . \tag{60}$$

The right hand side is approximated by inserting in place of $\phi(\boldsymbol{x})$ the expectation value $\dfrac{g}{m}\gamma(\boldsymbol{x})$ at $t=t^b$ (which is the exact valve in the limit $\dfrac{m}{g}\to 0$). We neglect still further the k-dependence in ω_k' and write $\omega_k'=\omega$. The last approximation will not be a bad one but we can take it as a model with $\kappa=0$. Then (53) is simplified as

$$\gamma(\boldsymbol{x}) = \int d\boldsymbol{x}\rho(\boldsymbol{x}-\boldsymbol{r}(t))\frac{\sin\omega(t^b-t)}{\omega} , \tag{61}$$

and we finally get

$$\dot{p}_b = \frac{g^2}{m}\int_{t^a}^{t^b} dt\int d\boldsymbol{x}\rho(\boldsymbol{x}-\boldsymbol{r}(t^b))\frac{\partial}{\partial\boldsymbol{x}}\rho(\boldsymbol{x}-\boldsymbol{r}(t))\frac{\sin\omega(t^b-t)}{\omega} . \tag{62}$$

Since $\rho(\boldsymbol{x})$ is assumed to be finite in the small region around $\boldsymbol{x}=0$, the term $\rho\nabla\rho$ has the appreciable value only near $t=t^b$. Then by looking at Fig.4, we conclude that $\rho\partial_z\rho<0$. Here we have assumed (56) and v is taken to be in the z-direction. Furthermore $\dfrac{\sin\omega(t^b-t)}{\omega}$ is positive near $t=t^b$. Therefore $\dot{p}_b$, directed along the z-axis, is negative and nearly independent of t^b. Consequently

$$\dot{p}_b = \frac{g^2}{m}\int_0^\infty dt d\boldsymbol{y}\rho(\boldsymbol{y}+\boldsymbol{v}t)\frac{\partial\rho(\boldsymbol{y})}{\partial\boldsymbol{y}}\frac{\sin\omega t}{\omega} \equiv Q \tag{63}$$

where $Q = (0, 0, Q_z)$ with $Q_z < 0$.

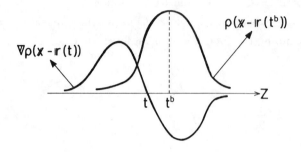

Fig.4 Typical form of $\rho(\boldsymbol{x} - \boldsymbol{r}(t^b))$ and
$\nabla\rho(\boldsymbol{x} - \boldsymbol{r}(t))$ appearing in (62)

Without the energy supply the particle loses energy. In order for (56) to be a consistent solution, the external field is applied by adding the Lagrangian,

$$L_\alpha = \alpha \cdot r, \quad \alpha = -Q. \tag{64}$$

This leads to the solution of the whole problem.

In the model above, we find that the macrovariables can be selected and be used as a detector variables in the following three cases;

a) $g/m \to \infty$: This has been discussed above. The macrovariable is $(m/g)\phi(\boldsymbol{x})$ and we have to supply the external field of the order g^2/m.

b) $T = t^b - t^a \to \infty$ but $g/m = finite$: It requires infinite time interval for observation but is a possible solution (Similar to Hepp (1972)). We keep the external field finite so that the time interval T should be sent to infinity in order to make Γ infinite, which is a necessary condition for obtaining the stationary phase.

c) $|\boldsymbol{v}|\to\infty$ and $g\to\infty$:   The external field is absent here. Since $|\boldsymbol{v}|$ is large the energy is supplied by the object particle. But for large $|\boldsymbol{v}|$, the probability of the interaction with $\phi(\boldsymbol{x})$ per unit length along the particle trajectory becomes small so that g should be large depending on $|\boldsymbol{v}|$. In this way we can make the size of the region large in the finite time interval where $\phi(\boldsymbol{x})$ takes appreciable finite value.

6. DISCUSSIONS

The purpose of the paper is to provide a unifying picture of the measurement in terms of the stationary phase in the action functional. Two simple models are given in order to illustrate the importance of the stationary conditions in the measurement theory. For the theory to be applicable to the real detector systems, more realistic Lagrangians have to be investigated of course. However the general principle given here will be operating also in a complicated system since the structure of the Hilbert space pointed out in Section 1 is common to any macroscopic systems.

Another problem is to study the non-ideal case where N is not strictly infinite. This is important both conceptually and practically. The neglected term of the order $1/\sqrt{N}$ is non-vanishing which may bring back the interference term after the measurement. The answer to this question is that in order to get the interference of the order unity after the measurement the macroscopic energy has to be supplied. The interference is lost unless such an energy is involved (Fukuda, 1989b). The issue is certainly connected with the theory of half or incomplete measurement. These points should further be studied.

References

Cini M 1983 Nuovo Cim. **73** 27

Feynman R P 1948 Rev. Mod. Phys. **20** 367

Feynman R P and Hibbs A R 1965 Quantum Mechanics and Path Integral (New York: McGraw-Hill)

Fukuda R 1987a Phys. Rev. **35A** 8

Fukuda R 1987b Phys. Rev. **36A** 3023

Fukuda R 1988 Quantum Mechanics of the Macroscopic System and the Theory of Measurement - Principle of Stationarity in the Action Functional - (Keio Univ. preprint)

Fukuda R 1989a Prog. Theor. Phys. **81** 34

Fukuda R 1989b Macrovariables and the Theory of Measurement (talk given at the 3rd Int. Symp. on the Foundations of Quantum Mechanics, in the Light of New Technology)

Hepp K 1972 Helv. Phys. Acta **45** 237

Namiki M, Ohnuki Y, Murayama Y and Nomura S 1986 *Proc. 2nd Int. Symp. on the Foundation of Quantum Mechanics, in the Light of New Technology (Tokyo: Phys. Socie. Japan)*

Quantum (Statistical) Mechanics, Measurement and Information

Roger BALIAN

INTRODUCTION

Ever since the beginning of quantum mechanics, the measurement problem has been a subject of sometimes discontinued but nevertheless recurrent concern. As students, most physicists have raised questions about it, but the spectacular and renewed successes of quantum theory were a strong incentive for them to work out applications without further worrying about foundations. In each generation, however, some dissatisfaction reemerges, which leads to slow progress, and we may expect that the measurement problem will some day be cleared up rather than forgotten. The situation recalls statistical mechanics, the foundations of which still give rise to scientific discussions, in spite of more than one century's achievements. Actually, the paradoxes of irreversibility and of Maxwell's demon share some common features with quantum measurements, to wit, they involve probabilities in a crucial fashion and their explanation requires elucidating the role of the observer.

We shall stress below such features, while we attempt to present a few simple ideas that may be unanimously accepted. Overlap with the current literature is thus unavoidable (see the review book by Wheeler and Zurek 1983). Some of the ideas gathered here have been developed in articles recently published, and we refer to these articles (listed at the end) for more details and references. We begin by reviewing the principles of quantum mechanics in a form suited to our purpose (section 1), then we discuss the differences between measurements and preparations (section 2) and the reduction of the state (section 3). The last sections are devoted to some aspects of quantum measurements connected with information theory. Setting aside the analysis of models of actual quantum measurement processes, we focus on general properties of idealized measurements.

1. PRINCIPLES OF QUANTUM MECHANICS

We adhere to the belief that, although quantum mechanics is intrinsically a *statistical* theory, it is *complete* in the sense that no underlying hidden variables should be looked for. Moreover, we accept the Bayesian viewpoint on probabilities in that they are not a property of the system as such, but rather a mathematical tool for making consistent and reasonable predictions on it from some previously acquired evidence (Cox 1946). Hence, since probabilities are unavoidable in quantum mechanics, we have to face from the start a partly anthropocentric character of this theory. In particular, even if we deal with an individual system, we should consider it as an element of a *statistical ensemble* of systems, all produced under similar conditions, but subject to individual variations in at least some of their measured properties. Quantum theory then appears as a means for making statistical predictions on this ensemble, or equivalently best guesses on one of its elements. The observer is thus part of the game, but neither more nor less than in any statistical theory: the probabilities characterize *his* knowledge of the system.

The measurements and preparations (in particular by means of a reduction of the state) are the processes through which information is acquired. They serve to confirm or modify the a *priori* probability distribution assigned to the system. Exact predictions or results should be regarded as limiting cases involving a unit probability.

1.1. Physical quantities

In this probabilistic view, physical quantities play the role of random variables. However, as a basic principle of quantum mechanics, they constitute a *non-commutative algebra*. Physical quantities are thus represented by *hermitean operators* or matrices in the Hilbert space associated with the system under consideration.

The hypothesis that *all* such operators can represent a physical quantity implies the superposition principle, as well as Pauli's symmetry or antisymmetry principle for indistinguishable particles. Otherwise, superselection rules arise.

It should be stressed that the specific features of quantum mechanics, such as the coherent superposition of state vectors, Heisenberg's or Bell's inequalities, all arise from the non-commutation of observables. Indeed, once the physical quantities are identified with the elements of some algebra, it is no longer necessary to postulate that quantum states are represented by vectors or wave functions. Instead, quantum states can then be introduced along the lines of algebraic theory, as we now recall (see, for instance, the books of Davies 1976, Piron 1976, Thirring 1981, Emch 1984); this approach to quantum mechanics is the same as in *classical* statistical mechanics, which is recovered if the algebra is commutative.

1.2. States

A state specifies our information at a given time on an ensemble of sytems: it allows us to make statistical predictions about any physical quantity. A quantum state is therefore identified with a correspondence *assigning to each operator its expectation value.* This correspondence is assumed to be linear, so that the expectation value of the sum of two physical quantities equals the sum of their expectations, even if they do not commute. Moreover, it should produce real fluctuations, i.e., the expectation value of the square of any hermitean operator should be non-negative.

This principle about quantum states amounts to defining them as *density operators* ρ or density matrices, which play the role of probability distributions (the proof, not simple for infinite dimensional algebras, is given by algebraic theory). The correspondence defining the expectation value of any observable A is thus implemented as

$$\langle A \rangle = \mathrm{Tr}\, \rho\, A \,, \tag{1}$$

an expression which summarizes all the predictions that can be made on the system. Measurements can modify the state of an ensemble of systems, not only by perturbing them, but also by changing our knowledge. Pure states appear as special limiting cases, for which the density operator reduces to a projector, but which still retain a probabilistic character.

A more conventional approach consists in postulating, conversely, that the "intrinsic states" of a system are represented by the vectors of a Hilbert space. Density operators are subsequently introduced to account for some lack of knowledge about the state vector. This viewpoint can be criticized for many reasons. Below we list a few of them in order to explain why we find it necessary to base the concept of "state" on density operators rather than on state vectors.

If correlations have developed between two systems, it is impossible in the language of state vectors to describe each one separately. The only consistent representation then relies on the use of a wave function for the overall system, and hence for the whole set of systems with which interactions took place in the past. However, a measurement process used as a preparation obviously requires to generate first correlations between the object and the apparatus, then to *separate* them from each other. Thus, asking questions about the object itself after this preparation *conceptually* requires the use of its density operator. In this respect, we disliked the title of the Rome Colloquium ("Quantum Theory without State Vector Collapse") from which the present book issued ... but deleting the last word of this title would make it agreeable!

If state vectors are chosen as the basic concept of states, randomness is introduced in two steps of apparently different natures. Each possible state vector includes an irreducible probabilistic aspect attributed to quantum mechanics, whereas the probabilities of the vectors which serve to build the density operator are regarded as accounting for our ignorance about the preparation. However, once state vectors and their probabilities have merged into a density operator, it is *no longer possible to disentangle* in a unique fashion this density operator into its original elements. For any non-pure state, no physical meaning can therefore be given to the analysis of a density operator in terms of underlying state vectors and their probabilities. To illustrate this point, consider an unpolarized spin. According to (1), the whole information available is included in its density operator (proportional to unity). However, if we wish to interpret this mixed state in terms of state vectors, we may regard it either as an incoherent superposition of spins pointing in arbitrary directions, or equivalently of spins polarized in the $+z$ and $-z$ directions with probabilities $1/2$, or of spins polarized in any other pair of opposite directions. This ambiguity indicates that state vectors are not an adequate description of the physical world (unless we allow ourselves to speak of the state vector of the whole universe); they can of course be used as mathematical tools, but not for conceptual purposes. Even if we build an unpolarized ensemble of spins by mixing at random two populations of spins previously prepared with the orientations $\pm z$, the description of the resulting state should not refer to these original orientations, since they are lost by the mixture. Indeed, nothing can distinguish this state from the one obtained by mixing spins polarized in the $\pm x$ directions. It is thus natural to identify the concept of state with a density operator best suited to describe the physical situation and not try to analyse it further. Otherwise, one might be led to think that privileged, although undetermined, directions preexist in the state of an unpolarized spin (or of a singlet pair in the EPR experiment), and that they become material after measurement. We feel some paradoxical aspects of quantum measurements arise from conscious or unconscious use of state vectors underlying the density operator, a language improper to suit quantum reality.

We shall not dwell further on the other conceptual advantages of density operators over state vectors (Wigner 1963, Balian 1989a, Ballentine 1990). Let us just mention that the present viewpoint is more directly related to experiment, more general (it covers partial information, random evolutions, macroscopic systems), and better adapted to deal with topics such as Liouville representations, the classical limit or quantum field theory. For completeness, we recall finally the last principle of quantum mechanics.

1.3. Evolution

The algebra of observables remains unchanged in time if the hamiltonian of the (isolated) system is known. This is expressed by the Heisenberg (or the Liouville-von Neumann) equation of motion. Indeed, the algebra is invariant under unitary transformations, and the hamiltonian is the generator of the unitary time-evolution.

2. MEASUREMENTS AND PREPARATIONS

At this point we find it useful to discuss the differences between the various types of processes that provide us with information about a system (Balian and Vénéroni 1987). By means of repeated operations on samples of an ensemble, both preparations and measurements lead to a better (statistical) knowledge of states. They are distinguished, however, by their *a priori* or *a posteriori* character, which brings in a lack of symmetry between them, especially when they are not complete.

Measurements aim at determining the density operator ρ, yet unknown, of a statistical ensemble of systems *previously produced* by some reproducible means. A *complete set of measurements* should determine all the matrix elements of ρ, e.g., all three polarizations $\langle \sigma_x \rangle$, $\langle \sigma_y \rangle$ and $\langle \sigma_z \rangle$ for a spin 1/2. Such a process requires several independent series of observations performed with different devices, since all the observables cannot be measured simultaneously.

Most often, only *incomplete measurements* are performed. They yield *some* data of the type (1), and are not sufficient to determine ρ fully — except (i) if the unknown ρ happens to have been a pure state, and (ii) if by chance the apparatus was designed to detect that very state (for instance, if a spin 1/2 polarized in the $+x$ direction goes through an analyser oriented along $\pm x$). Otherwise, many density operators ρ are compatible with the known expectation values, and a statistical inference procedure is required in order to assign a well-defined density operator to the state.

Among incomplete measurements, a class which we call *diagonally complete measurements* is of particular interest. They are the most informative measurements using a single device. They are defined as the ones which determine simultaneously a complete set of commuting observables. They yield all the diagonal elements of ρ in the common eigenbasis of these observables, but leave the off-diagonal elements unspecified. Equivalently, the measurement of a single observable having a non-degenerate spectrum is diagonally complete (for instance, the measurement of the component of a spin 1/2 in some given direction).

In this paper, the terms measurements and observables refer to processes determining expectation values of the type (1) of physical quantities A attached *to the system only*. More general measurements may be imagined, which are represented by positive operator-valued measures on density operators, and of which the correspondence (1) is just a special case. For instance, Davies (1976, p.16) discusses successive Stern-Gerlach experiments first in the $\pm z$ direction then in the $\pm x$ direction. Whether or not the particles are selected according to their trajectory after the first splitting, the statistics on σ_x cannot be represented by (1). Actually, this process should be identified as a measurement of σ_x at a time t_2 preceded by depolarization in the x-direction at a time t_1. If we refer to the initial time t_0 at which the spin was in the state ρ, the quantity thus measured, namely σ_x carried back from t_2 to t_0 by means of the Heisenberg equation, depends not only on the object, but also on the depolarizing device. Such processes are therefore *imperfect mesurements*, which give us information on physical quantities pertaining not only to the system alone, but also to part of the apparatus. The irreversible process that precedes the measurement of σ_x is the result of a coupling with degrees of freedom not described by ρ.

Preparations are often achieved by *filtering* the results of some special type of incomplete measurements, named by Pauli measurements of the first kind. We refer to them as *ideal measurements* and now list their properties. Of course, an ideal measurement should not destroy the object. It should moreover leave it in some definite final state which is not necessarily the same as the state ρ before measurement. We can distinguish in this respect two types of final states after an ideal measurement of the observable A :

(i) The *whole statistical ensemble* of systems that have undergone the experiment is retained, irrespective of the observed result. We denote as ρ_A the density operator representing this state.

(ii) After registration of the results, informing us about the eigenvalues a_α of A, we *select the subensemble* of outcoming systems for which some specific value a_α has been observed. We denote as ρ_α the density operator describing the state of this subpopulation. Such a filtering is the process which *prepares the state ρ_α*.

To be useful as preparations, ideal measurements should also have the following properties:

(a) *Repeatability*. The outcoming filtered state ρ_α should be such that a new measurement of A *again yields the answer* a_α. This property establishes a correlation between successive states of *individual systems*, irrespective of the original statistical ensemble before measurement. This correlation is complete, in spite of the dispersion of the eigenvalues of A which existed in the original state ρ of the whole ensemble.

(b) *Compatibility*. If two observables A and B *commute*, the ideal measurement of A may be supplemented (at least in thought) by a measurement of B without any perturbation on the statistical results of either. This measurement of A does not affect the predictions about B, provided the whole original ensemble is kept. Hence, in the unsorted final state ρ_A, the statistics of B is the same as in the initial state, a property expressed by

$$\text{Tr } \rho \, B = \text{Tr } \rho_A B \qquad (2)$$

for any B such that $[A, B] = 0$. (Note that A and B may be correlated so that $\text{Tr } \rho_\alpha B$ depends on α, e.g., for $B = A^2$.)

We regard these properties as *defining an ideal measurement*. A genuine theory of measurement should analyse the dynamics of the coupled system formed by the apparatus plus the object and prove that (a) and (b) can be satisfied at least on models simulating actual preparation processes.

If a measurement of A is *both ideal and diagonally complete*, then the eigenvalues a_α are non-degenerate, and filtering constitutes an *ideal preparation*. The outcoming state ρ_α is then *pure*. This explains the prominent role of diagonally complete measurements in the literature. As preparations, they allow us to *predict* with certainty the state produced; but as measurements, they are incomplete since they are not sufficient for *retrodicting* the off-diagonal elements of ρ.

In an *ideal but incomplete measurement* of A, selection of an eigenvalue a_α is in general not sufficient to single out a pure state. The resulting preparation is *not ideal*: the outcoming filtered state ρ_α is not pure, and it *depends on the original state* ρ. Its expression is given by the "collapse" or "reduction" of this initial state ρ.

The interrelations between the various types of measurements and preparations are therefore somewhat intricate. There is no symmetry between information gained on a preexisting state and information serving to prepare a new state (pure or not). In particular, a diagonally complete ideal measurement *tells us nothing* about the off-diagonal elements of ρ, but it *fully determines* the prepared states ρ_α. Because this lack of symmetry arises from non-diagonal elements of the density matrix, it is a specific feature of quantum mechanics.

3. REDUCTION OF A STATE BY AN IDEAL MEASUREMENT

Consider an ideal measurement of the observable A, used as a preparation. Let us denote as π_α the projector on the subspace of the Hilbert space characterized by the eigenvalue a_α of A. The reduction of the original state ρ by this measurement expresses the final state ρ_A of the system, after interaction with the measuring apparatus and separation from it, as

$$\rho_A = \sum_\alpha \pi_\alpha \, \rho \, \pi_\alpha \,. \tag{3}$$

If moreover the result a_α of the measurement is selected, ρ is reduced to the final state

$$\rho_\alpha = \frac{\pi_\alpha \, \rho \, \pi_\alpha}{\mathrm{Tr} \, \rho \, \pi_\alpha} \tag{4}$$

which describes the resulting subensemble.

These properties are sometimes posed as an additional principle of quantum mechanics. It is interesting to note, however, that they follow from the contents of sections 1 and 2. A formal proof is straightforward (Balian 1989a). Let us just sketch here its main steps and the premises on which they rely. A first ingredient is the *repeatability* of the measurement, defined in section 2 (a). It is expressed by

$$\mathrm{Tr} \, \rho_\alpha \, \pi_\alpha = 1 \,, \tag{5}$$

which means that the probability to get a_α in the state ρ_α is 1. Taking into account the *positivity* of the density operator ρ_α, the condition (5) implies that, in a basis which diagonalizes A, ρ_α includes only the diagonal block α, i.e.,

$$\rho_\alpha = \pi_\alpha \rho_\alpha \pi_\alpha \ . \tag{6}$$

The overall outcoming ensemble is split by sorting the results a_α into subensembles described by ρ_α and containing each a proportion $\mathrm{Tr}\,\rho\,\pi_\alpha$ of systems. This is expressed by

$$\rho_A = \sum_\alpha \rho_\alpha \mathrm{Tr}\,\rho\,\pi_\alpha \ . \tag{7}$$

The proof is achieved by using the *compatibility* condition (2), which means that ρ_A and ρ have the same diagonal blocks. Together with (6) and (7), this determines ρ_A and ρ_α as (3) and (4), respectively.

If the measurement of A is not only ideal but also diagonally complete, (4) implies that ρ_α is just the projector π_α on the single eigenvector of A associated with a_α. Otherwise, ρ_α and ρ_A obviously depend on ρ; the compatibility property is then essential to determine them.

The two expressions (3) and (4) clearly exhibit the origin of the changes undergone by the state of the system in an ideal measurement of A. The change from ρ to ρ_A is the result of two facts: (i) The evolution of the coupled system formed by the object plus the apparatus (which is in the state ρ' before interaction) leads them from the initial state $\rho \otimes \rho'$ to a state ρ_{tot} involving correlations between object and apparatus. This state ρ_{tot} refers to the overall ensemble. (ii) Because we are interested in the state of the object only, we eliminate the whole information about the apparatus by performing a partial trace $\mathrm{tr}_{\mathrm{ap}}$ over it, which defines

$$\rho_A = \mathrm{tr}_{\mathrm{ap}}\rho_{\mathrm{tot}} \ . \tag{8}$$

Thus, ρ_A is the result of *interaction* then *separation*. Any process involving these two features modifies ρ into (8), and this change is non-unitary because some degrees of freedom are eliminated. The fact that we are dealing with an ideal measurement appears in the specific form (3) obtained for (8). On the one hand, the disappearance of the off-diagonal blocks is needed to distinguish later on the eigenvalues a_α from one another. On the other hand, the invariance of the diagonal blocks (a consequence of compatibility) means that an ideal measurement modifies the system as little as possible.

The density operator ρ_A disregards *all* the correlations between object and apparatus that have been produced by the interaction process. Part of these correlations are, however, the essential feature of a measurement. Indeed, a measuring apparatus involves a *registration device*, which played no role above, but which governs both the gain of information about the original state ρ (through the determination of $\mathrm{Tr}\,\rho\,\pi_\alpha$) and, if the measurement is ideal, the preparation of the final states ρ_α through selection. The registration involves a macroscopic variable that behaves classically and takes discrete values a'_α, the statistical fluctuations of which are small compared to their differences. These values a'_α are fully correlated with a_α, i.e., the occurrence of a'_α means that A takes the value a_α in the final state ρ_A. Therefore, each a'_α is associated with the single term α of the sum (3) or (7), and the *selection of the result a'_α* produces a subensemble in which the system is in the state (4). The normalization factor accounts for the number of elements of this subensemble.

Only the process leading from ρ to ρ_A is specific to quantum mechanics. In classical theory, off-diagonal elements are absent and ρ_A does not differ from ρ. On the other hand, the splitting of the original ensemble into subensembles α according to the outcome of the measurement and the resulting reduction of ρ_A into ρ_α are exactly the same as in *any statistical problem*: the systems are prepared in states ρ_α better defined than before by taking advantage of the information gained through observation (although in quantum theory ρ_α always remains probabilistic even if it is pure). This idea will be made quantitative in section 5.

4. INFORMATION AND ENTROPY IN MEASUREMENTS

Since a measurement is aimed at gaining information, it is natural to estimate its efficiency by evaluating the *amount of information* that it provides. Information theory associates with any probability distribution p_m its statistical entropy

$$S(p_m) \equiv -\sum_m p_m \ln p_m, \tag{9}$$

which measures the *amount of uncertainty*, or of missing information, due to the statistical nature of our knowledge. Its extension to quantum theory, the *von Neumann entropy*

$$S(\rho) \equiv -\mathrm{Tr}\, \rho \ln \rho, \tag{10}$$

describes likewise the randomness of the state ρ; it vanishes for the pure states, which are the best specified ones.

Several statistical entropies are introduced in the context of an ideal measurement of the observable A, and they should not be confused (Balian, Vénéroni and Balazs 1986). By focusing on the probabilities

$$\mathrm{Tr}\, \rho \pi_\alpha = p_\alpha \tag{11}$$

of the various outcomes a_α of the experiment, one is led to define the *uncertainty of A in the state ρ*,

$$S(A/\rho) = -\sum_\alpha p_\alpha \ln p_\alpha, \tag{12}$$

also termed A-entropy or measurement entropy. It can be interpreted as the information which is missing when only the probabilities p_α are known, each event α having the same a *priori* weight.

However, in case the eigenvalues a_α have the degeneracies g_α, it is natural to use another definition, in which each event α is weighted by g_α, so as to account better for the structure of the Hilbert space and of the observable A. Equivalently, we introduce the state

$$\rho_{0A} \equiv \sum_\alpha q_\alpha \pi_\alpha \equiv \sum_\alpha \frac{p_\alpha}{g_\alpha} \pi_\alpha, \tag{13}$$

for which each eigenvector of A has the probability $q_\alpha = p_\alpha / g_\alpha$. This leads us to define a statistical entropy different from (12) but also associated with the measurement of A, namely

$$S(\rho_{0A}) = -\mathrm{Tr}\ \rho_{0A}\ln\ \rho_{0A} = -\sum_\alpha g_\alpha q_\alpha \ln\ q_\alpha. \tag{14}$$

It is easily shown that ρ_{0A} is the state that leads to the largest value of the von Neumann entropy (10), within the class of states providing the same statistics (11) as ρ for the observable A. Accordingly, (14) is identified with the *relevant entropy* (Balian, Alhassid and Reinhardt 1986) *of ρ relative to the measurement of A*. The states ρ and ρ_{0A} are equivalent with respect to this measurement, but ρ_{0A} contains the *least possible amount of information* required to account for the data (11). The two entropies (12) and (14) are related by

$$S(\rho_{0A}) = S(A/\rho) + \sum_\alpha p_\alpha \ln\ g_\alpha. \tag{15}$$

It is also natural to characterize the information contents of the states ρ and ρ_A by means of (10). They satisfy the inequalities

$$S(\rho) \leq S(\rho_A) \leq S(\rho_{0A}), \tag{16}$$

to which we return in section 5, where we discuss the amount of information gained through measurement of A.

Another aspect of information theory relevant to incomplete measurements and preparations refers to the *assignment of a state* to some system on which only *partial knowledge* is available. A completely specified (pure) state is produced by an ideal preparation; likewise, owing to the reduction (4), a preparation based on an ideal but not diagonally complete measurement leads to a state ρ_α which is entirely determined if the original state ρ is known. However, other incomplete preparation processes, as well as all incomplete measurements (as defined in section 2), are not sufficient to determine fully the density operator. Some data of the form (1) provide constraints on its matrix elements, but a *statistical inference* procedure is required to achieve the assignment.

To this end, it is natural to resort to the *maximum entropy criterion*, which consists in looking for the maximum of the von Neumann entropy (10) subject to the constraints brought in by the known data (Jaynes 1957). It means that, among all states which account for these data, the *least biased* state is the one which contains the *smallest amount of information*. The maximum value of S is then interpreted as the uncertainty associated with the available data. The maximum entropy criterion can be *derived* in quantum mechanics (Balian and Balazs 1987) by relying on the principle of indifference, which states that equal probabilities should be assigned to a set of events all placed on the same footing. A special case has implicitly been considered just above, since (13) is the state that should be assigned to a system if only the probabilities p_α of the eigenvalues a_α of A are known.

The above considerations refer to entropy in the sense of information theory. Entropy in the sense of *statistical mechanics* or thermodynamics is also relevant to the analysis of measurement processes. Indeed, the registration of the outcomes a'_α (in one-to-one correspondence with the eigenvalues a_α of A) is an *irreversible* process, *triggered* by the interaction of the observed object with the detector. This measuring device must have small statistical fluctuations, so that the outcomes a'_α are distinguished clearly from one another, a property achieved in general by its macroscopic size. Its study therefore pertains to the statistical mechanics of irreversible processes, with the standard difficulties of this theory; additional problems arise from the existence of several possible final states a'_α and from the role of the object in obtaining such or such a'_α. Bennett (1982) has emphasized the need for the initial state ρ' of the apparatus to be metastable and to have some minimum entropy, because it should be able to evolve towards one among several possible final states (labelled by a'_α) without any bias on the outcome.

As a matter of fact, the registration device of a measuring apparatus is analogous to a physical system having *several phases*, which are equivalent and can be distinguished from one another by the value of some *order parameter* a'_α. (For instance, in a bubble chamber, the position of the bubbles is directly related to the order parameter of the gas-liquid transition.) Although such a system has an underlying quantum nature as any physical object, its order parameter behaves *classically* in the following sense: starting from an initial state chosen at random, the evolution leads to an equilibrium situation where a'_α has a well-defined value; one *never* observes the formation of a state which would involve a *coherent superposition* of states associated with different values of the order parameter. (For instance, each domain in a ferromagnet has a magnetization oriented in some direction, and coherent s-wave superpositions, although not forbidden, are not seen.) A detector has the same properties, and in addition its final state a'_α is determined statistically by its interaction with the measured quantum object. The well-known irreversibility of quantum measurements is thus directly related to this type of irreversibility in statistical mechanics. We hope to see some progress along these lines in the future.

5. BALANCE OF INFORMATION IN AN IDEAL MEASUREMENT

Discarding from now on the apparatus, let us compare the entropies of the states of the object before and after the measurement of A. Several inequalities can easily be proved, most of them well-known with interesting interpretations (Balian 1989b). The first inequality (16) means that an ideal measurement of A produces an increase $S(\rho_A) - S(\rho)$ of the entropy of the state of the object (even if there is no registration or selection of the results). In other words, a price has to be paid for a measurement of A, since it is necessarily accompanied by a *destruction of our information* about the physical quantities incompatible with A. This loss is associated with the truncation of the off-diagonal blocks of ρ. It arises from the interactions of the object with the measuring apparatus, which create the correlations used later on to prepare the states ρ_α. Actually, the process leading from the initial state $\rho \otimes \rho'$ of the overall system (apparatus plus object) to ρ_{tot} might be reversible, but the increase of entropy of the object itself is due to the partial trace (8) which discards all the correlations between this object and the apparatus.

Moreover, among all the states satisfying the compatibility conditions (2), ρ_A turns out to be the one which has the largest entropy. Hence, ρ_A not only *retains the whole information about the observables commuting with* A, but also contains *no more information* (in contrast to the original state ρ). The loss of information $S(\rho_A) - S(\rho)$ has the largest possible size compatible with the retained data (2). The repeatability of the measurement of A thus produces the *same result* ρ_A *as the maximum entropy criterion*, and the entropy $S(\rho_A)$ of the reduced state is identified with the relevant entropy of ρ relative to all observables commuting with A.

While *all* the information about the observables that *do not* commute with A is lost in ρ_A, the second inequality (16) states that still less information is contained in ρ_{0A}. Indeed, ρ_{0A} was defined by the maximum entropy criterion as the state which accounts *only* for the statistics of A, and $S(\rho_{0A})$ was identified with the relevant entropy of ρ relative to A.

The entropy of a filtered state ρ_α measures the uncertainty remaining about the observables other than A when the result a_α has been selected. Its mean value

$$\langle S(\rho_\alpha) \rangle \equiv \sum_\alpha p_\alpha S(\rho_\alpha) \tag{17}$$

satisfies several relations having natural interpretations. The identity

$$S(\rho_A) = S(A/\rho) + \langle S(\rho_\alpha) \rangle \tag{18}$$

means that the overall uncertainty about the outcoming ensemble (after measurement) is the sum of the uncertainty about A and of the *average uncertainty remaining* once A is known. It is a special case of *additivity of information*. It can be also interpreted as expressing that *information can be transformed into negentropy:* the information

$S(A/\rho)$ acquired by registering the outcomes a'_α of the experiment can be used to sort the systems according to a_α and thus to lower their average entropy in the ensemble from $S(\rho_A)$ to $\langle S(\rho_\alpha) \rangle$. (Conversely, mixing the various populations ρ_α raises the entropy by an amount equal to the information lost about A.) A special case of (18) is (15), since $S(\rho_\alpha)$ is equal to $\ln g_\alpha$ when ρ_A reduces to ρ_{0A} and ρ_α to π_α/g_α.

As a consequence of (16) and (18), we have

$$S(\rho) \leq S(A/\rho) + \langle S(\rho_\alpha) \rangle , \qquad (19)$$

which means that quantum mechanics spoils the additivity of information. The *total uncertainty* about A and about the other observables *increases* through the measurement process. The gain of information about A is *paid for by the destruction* of some other information.

Finally, the inequality (Lindblad 1973)

$$S(\rho) \geq \langle S(\rho_\alpha) \rangle = S(\rho_A) - S(A/\rho) \qquad (20)$$

means that an ideal measurement including selection of the results *lowers the entropy* of the objects (at the expense of the apparatus and owing to the registered information), in spite of the unavoidable quantum irreversibility (19). The equality in (20) holds only if all $S(\rho_\alpha)$ are equal to $S(\rho)$. In the special case when ρ is a pure state, (20) implies that all ρ_α are also pure, a property obvious from (4). In the general case, this inequality has another interesting interpretation: although ideal, the measurement of A destroys the entire information $S(\rho_A) - S(\rho)$ about the observables incompatible with A; but *this quantum loss is smaller than the average gain* $S(A/\rho)$ of information about A that results from the observation of the result.

Acknowledgments

I wish to thank Alan Macdonald for interesting discussions, as well as John Palmeri, Jean-Marie Normand and Gilbert Mahoux for a critical reading of the manuscript.

References

Balian R 1989a *Am. J. Phys.* **57** 1019
Balian R 1989b *Eur. J. Phys.* **10** 208
Balian R, Alhassid Y and Reinhardt H 1986 *Phys. Reports* **131** 1
Balian R and Balazs N L 1987 *Ann. Phys.* **179** 97

Balian R and Vénéroni M 1987 *Ann. Phys.* **174** 229

Balian R, Vénéroni M and Balazs N L 1986 *Europhys. Lett.* **1** 1

Ballentine L E 1990 *Quantum Mechanics* (Englewood Cliffs: Prentice Hall)

Bennett C H 1982 *Int. J. Theor. Phys.* **21** 905

Cox R T 1946 *Am. J. Phys.* **14** 1

Davies E B 1976 *Quantum Theory of Open Systems* (London: Academic)

Emch G G 1984 *Mathematical and Conceptual Foundations of 20th-Century Physics* (Amsterdam: North-Holland)

Jaynes E T 1957 *Phys. Rev.* **106** 620, **108** 171

Lindblad G 1973 *Comm. Math. Phys.* **33** 305

Piron C 1976 *Foundations of Quantum Physics* (Reading: Benjamin)

Thirring W 1981 *A Course in Mathematical Physics* (New York: Springer)

Wheeler J A and Zurek W H, eds 1983 *Quantum Theory and Measurement* (Princeton: Princeton University)

Wigner E 1963 *Am. J. Phys.* **31** 6

Further references are included in the above articles and books.

State Vector Collapse as a Classical Statistical Effect of Measurement

Marcello Cini and *Maurizio Serva*

1. INTRODUCTION

The absence of a general consensus among physicists about the old question of the measurement induced state vector collapse in quantum mechanics is, at first sight, puzzling. It is true, of course that the question cannot, at least as yet, be settled by submitting it to the final decision of experiment. Still, one would like to understand better the main issue of disagreement which leads to such a wide variety of possible solutions. We believe that the origin of this disagreement lies in a fundamental ambiguity which exists in our understanding of the relation between quantum mechanics (QM) and classical statistical mechanics (CSM). For this reason we will start by discussing at length this ambiguity, whose clarification is, in our opinion, preliminary to any attempt to find a solution of the question which is the central theme of our Colloquium. We will come back to this question, therefore, only in the final part of this talk.

We will start by asking a very simple question. A basic axiom of QM is that, if the state of a particle is the superposition of two states belonging to different values of a given dynamical variable, it is only in the act of measurement that the variable acquires, at random, one of these two values. On the other hand classical mechanics assumes that a variable has always a precise value whether we measure it or not. One may ask therefore how is it that, in the limit when QM tends to CSM, the statement that a given variable of a physical system has always a precise value independently of having been measured or not - a meaningless statement in QM - gradually becomes meaningful. In other words, how can it be that QM, which is a theory describing the intrinsically probabilistic properties of quantum objects, becomes, in this limit, a statistical theory describing a probabilistic knowledge of intrinsically well determined properties of classical objects?

The first thing to do, of course, is to define properly what we mean by the "limit when QM tends to CM". This procedure is not as simple as letting h go to zero. In fact h is a constant of nature and one should rather take the correct limit for the appropriate dynamical variables when they become large compared with the relevant atomic units. To this purpose, however, one should realize that there are at least two different cases in which one expects that QM should approach CM.

The first case refers to the limit of large "quantum numbers" for a given quantum system, in the sense of Bohr's correspondence principle. This is the limit in which the variation ΔS of the phase of the wave function within the region where it is substantially different from zero is much larger than h (in other words the limit in which the wave length of a wave packet is much smaller then its extension), or, if the state is stationary, the energy E_n is much larger than the ground state energy E_o. The question is now: when the particle is in a state of this kind can one still maintain that its variables are not defined before they have been measured? Or, more precisely, if the state vector of the particle is a superposition of two states corresponding to macroscopically different values of a given variable, can one still maintain that this variable acquires, at random, one of these values only when it is measured? This would be in contrast with the classical statistical picture which supposes that the macroscopic variable does have one of the two possible values independently of whether it has been measured or not.

The second one, the case of a macroscopic body, is even more puzzling. The system is now made of an enormous number N of elementary quantum systems and has a correspondingly large number of degrees of freedom. For such a system it is generally possible to define at least a pair of collective (pseudo)conjugate variables (e.g. the center of mass coordinate and its velocity) who satisfy two conditions: (a) the commutator of these collective variables vanishes as N goes to infinity; (b) they are decoupled from the variables of the individual particles, and their Heisenberg equations of motion tend, in this limit, to the classical equations of motion.

One might think therefore that in this case property (b) gives an answer to our question. Since the classical equations of motion can be solved with precisely given initial values for both these variables their value is completely determined at all later times. This means that these variables "have" precisely determined values whether they are measured or not.

This statement is, however, only a partial answer, because it is true only for the particular choice of initial conditions just considered. In fact, QM yields, as we shall see in more detail in a moment, a statistical distribution in the phase space of the collective variables that does not generally reduce to a single point. It may happen, for example, that a pure quantum mechanical state corresponds to a limiting classical statistical distribution having the form of two delta functions centered on two different classical trajectories. Do we have, here again, to assume that our macroscopic body choses, at random, one of the two possible trajectories (on each of which the center of mass position and velocity are defined at any time), only when a measurement is performed on it? Should we not rather assume, in view of the fact that the body is macroscopic, that the right description is given by CSM, and that the double delta function simply reflects our ignorance?

In the following we will discuss this puzzling question showing that if we choose the second answer, consistently with the physical content of CM, we are led also to admit that the widely spread belief in the current interpretation of QM needs revisiting. If we do not wish to do so we have to adopt a strictly subjective conception of the classical world outside us, implying that also macroscopic objects are not localized anywhere before we look at them.

2. THE CLASSICAL LIMIT OF A QUANTUM STATE

We shall now discuss the statistical properties of a quantum state in the classical limit. As is well known, Wigner (1932) has defined a quantum mechanical function $W(x,p)$ by means of the relation (for simplicity we put $h = 1$, but we will later introduce, when necessary, explicitly h):

$$W(x,p) = \pi^{-1} \int \psi^*(x+y) \, \psi(x-y) \, \exp(2ipy) \, dy \tag{1}$$

This function may be used to compute the expectation value of any quantum variable $\mathbf{A(x,p)}$ function of the operators $\mathbf{x}$ and $\mathbf{p}$, by means of an expression:

$$<\mathbf{A}> \equiv \int \psi^*(x) \, \mathbf{A}(x,-i\,\partial/\partial x) \, \psi(x) \, dx = \iint W(x,p) \, A(x,p) \, dx \, dp \tag{2}$$

provided one takes

$$A(x,p) = 2 \int <x+z|\mathbf{A}|x-z> \exp(-2ipz) \, dz. \tag{3}$$

W(x,p) corresponds to the classical distribution function f(x,p) in phase space because eq. (2) is formally identical to the classical expression

$$<A> = \iint dx \, dp \, A(x,p) \, f(x,p). \tag{4}$$

where A(x,p) is the classical variable which corresponds to $\mathbf{A(x,p)}$.

Of course, W(x,p) is not everywhere positive as the classical f(x,p) and therefore can not be interpreted as a distribution function. Nevertheless eq.(2) reduces, generally, to eq.(4) in the classical limit. In the second part of this section we will show with a few examples that this limit consists essentially in considering large quantum numbers.

On a formal level one can simply suppose that h is negligibly small($h \rightarrow 0$). In this case $A(x,p) \rightarrow A(x,p)$, when the quantum variable $\mathbf{A(x,p)}$ has the same form of the classical variable A(x,p) because its matrix elements $<x|\mathbf{A}|p>$ differ from it only by terms of order h arising from the reordering of the operators $\mathbf{x}$ and $\mathbf{p}$. On the other hand it is well known that the time evolution of W(x,p;t) reduces to the Liouville equation

$$\partial W(x,p;t)/\partial t = - (p/m) \, \partial W(x,p;t)/\partial x + (\partial V(x)/\partial x)(\partial W(x,p;t)/\partial p) \tag{5}$$

when the potential V(x) is slowly variable.

This reduction is a necessary step in order to obtain the classical distribution in the limit $h \rightarrow 0$. One needs in fact:

$$\int_A dx \, dp \, W(x,p) \quad \rightarrow \quad \int_A dx \, dp \, f(x,p)$$

for a finite region A of the phase space. This limit obviously holds when W(x,p) tends to the corresponding f(x,p).

An interesting particular case is that of a free particle wave packet initially concentrated in a Gaussian region:

$$\psi(x,0) = (\pi\alpha)^{-1/4} \exp\{-(x-x_0)^2/2\alpha\} \tag{6}$$

The energy expectation value ε is given by

$$\varepsilon = h^2/4\alpha m \tag{7}$$

and the proper classical limit is provided by the limit $\varepsilon t \gg h$. Now one has:

$$W(x,p;t) \rightarrow (4\varepsilon t^2/m)^{-1/2} \exp[-m(x-x_0)^2/4\varepsilon t^2/ \delta[p-(x-x_0)m/t] \tag{8}$$

This is the classical distribution function $f(x,p;t)$ in phase space of a free Hamiltonian fluid concentrated initially (t=0) at $x=x_0$.

When the state is stationary with energy $E \gg E_0$ (ground state energy) one has

$$W(x,p) \rightarrow N\, p_E^{-1}[\, \delta(p-p_E) + \delta(p+p_E)] = N\, \delta(H-E) \tag{9}$$

where

$$H = (p^2/2m) + V(x) \tag{10}$$

and N is the phase space volume on the energy shell. Eq.(9) shows that the statistical properties of the quantum mechanical density matrix for a given energy E tend to those of the corresponding microcanonical ensemble of classical statistical mechanics.

Let us now see what happens when the state is a superposition

$$\psi = c_1\psi_1 + c_2\psi_2. \tag{11}$$

The total Wigner function will be the sum of the two Wigner functions of ψ_1 and ψ_2, weighted with the respective probabilities, plus an interference term, whose general features will be better understood by considering two complementary cases.

The first one is when the two wave functions are localized in two separate space regions. Then the contribution to $W(x,p;t)$ of the interference term contains a factor

$$\cos[p(x_1-x_2)/h] \tag{12}$$

where x_1 and x_2 are the mean values of x in ψ_1 and ψ_2. The contribution of this term to the expectation value of any variable $A(x,p)$ vanishes unless its p-dependence shows the same rapidly oscillating behaviour of the cosine factor. This is certainly not the case for the quantum variables considered here that have a classical limit.

The complementary case obtains when the two wave functions are not localized in separate space regions, but are labeled with two widely different values of the energy $E_1 \gg E_2 \gg E_0$ (ground state energy). Here again the interference term gives a vanishing contribution to expectation values of variables which have a classical limit. This can be easily seen in the simple example of a particle constrained between two fixed boundaries in which the interference term to the Wigner function oscillates with a factor

$$\cos\{[\sqrt{(2m\,E_1)} - \sqrt{(2m\,E_2)}]x/h\}. \tag{13}$$

Here the contribution of this term to the expectation value of a variable $A(x,p)$ is negligibly small unless its x-dependence shows the same rapidly oscillating behaviour of the cosine factor, a property which we do not expect a variable with a classical limit to possess.

It is therefore clear that the classical statistical ensemble corresponding to the quantum state (11) is always simply the union of the two classical statistical ensembles corresponding to the individual states, weighted with probabilities $|c_1|^2$ and $|c_2|^2$.

The case of a macroscopic body made of N particles coupled to each other may be treated by means of simple models and leads to exactly the same conclusions. The wave function of the system may be written in the form

$$\Psi = \phi(q)\,\chi\,(q_1, \dots q_{N-1}) \tag{14}$$

where ϕ is the center of mass wave function and χ is the wave function of the N-1 independent relative coordinates.

Consider for the moment the q-dependent part of ψ. If we take for ϕ the form (6), the classical limit of its Wigner function, of the form (8), is now attained as $N \to \infty$. We might instead consider however, any other wave function of the center of mass. A stationary state of energy $E \gg E_0$ would, for example, lead to a Wigner function which tends to (9) as $N \to \infty$. In other words the statistical properties of $\phi(q)$ are the same as those discussed in the previous section in the corresponding classical limit. The presence of the remaining variables does however make a difference in the case of a macroscopic body.

To see how this comes out let us consider a state described by a superposition of two wave functions of the form (14)

$$\Psi = c_1\phi_1\chi_1 + c_2\phi_2\chi_2 \tag{15}$$

The proper Wigner function of the intensive center of mass variables q,v is now constructed by integrating over all the microscopic variables. This leads to cross terms which are not only small for the reasons that made small the interference contribution (12) to the Wigner function derived from the wave function (11), but also because it is extremely unlikely that χ_1 and χ_2 are exactly the same. In general χ_1 and χ_2 are very different microscopically, namely orthogonal, even if they are macroscopically equivalent, and the cross terms actually vanish.

We may conclude therefore that there is always a one-to-one correspondence between a quantum mechanical state and a statistical distribution in phase space of classical statistical mechanics both in the case of a microscopic system and of a macroscopic body. The difference is that in the former the classical limit is attained when a suitable action variable is >> h, while in the latter the limit N→∞ is sufficient to ensure that the center of mass variables always behave as classical. We shall now discuss the consequences of this correspondence.

3. STATISTICAL PROPERTIES OF A QUANTUM STATE

It is now easy to discuss the statistical properties of a quantum mechanical state in the classical limit. It is particularly interesting to see what happens to the uncertainty product $\Delta x \; \Delta p$. For the free wave packet of eq.(6) one has:

$$(\Delta p)^2 = h^2/\alpha = 2\,\varepsilon m \tag{16}$$

$$(\Delta x)^2 = 2\,\varepsilon t^2/m + h^2/8\varepsilon m = (\Delta p)^2 t^2/m^2 + h^2/4(\Delta p)^2 \tag{17}$$

$$(\Delta x \, \Delta p)^2 = 4\,\varepsilon^2 t^2 + h^2/4 \rightarrow (2\,\varepsilon\,t)^2 \quad \text{for } \varepsilon\,t >> h \tag{18}$$

Eq.(17) resembles closely an old relation which marked a turning point in the history of physics: Einstein's formula for the energy fluctuation of radiation at thermal equilibrium expressed as the sum of two terms of different origin (Einstein 1909). In the case of radiation the quantum term arises from its particlelike properties and the classical term from the wavelike ones. In quantum mechanics the reverse happens. In our case the first (particle) term has a classical origin and the second (wave) term a quantum one. This separation, however, has been forgotten since the adoption of the standard interpretation of QM, which considers the fluctuations of the quantum variables as wholly due to their intrinsically undetermined nature. What we propose, on the contrary, is to take seriously this separation as physically meaningful. From this point of view, eq.(17) means that the spread of a quantum wave packet for large values of εt does not arise from an ontologically intrinsic delocalization of the particle, but, as it happens for classical particles, is a trivial consequence of the fact that the region where a particle may be found increases with time if its momentum is not precisely determined.

Stated differently, eq. (18) indicates that the really intrinsic quantum indeterminacy, reflecting the impossibility of simultaneous existence of position and momentum, is always the minimum one implied by the Heisenberg principle. Higher indeterminacies are instead of statistical nature, reflecting the actual displacement in space of particles with different momenta,and they survive in the classical limit. This picture is particularly relevant for the interpretation of the superposition (11) when the two states are macroscopically different. One can no longer say that only when a variable is measured it assumes a value within one of the two ranges allowed by either ψ_1 or ψ_2.

Since the statistical ensemble described by the density matrix of the state (11) is, in the limit of large values of the action, the weighted union of two disconnected classical phase space distributions, we are almost forced to conclude that it is meaningful to say that the particle belonged to one or the other distribution even before the measure had taken place. We will explain and justify this statement in the following sections, showing that it is indeed possible to give a more precise meaning to the separation between quantum and classical indeterminacy.

4. ALTERNATIVE INTERPRETATIONS OF CLASSICAL STATISTICAL MECHANICS

Assume now that a classical distribution function in phase space is at $t = 0$ of the form:

$$f(q,p;0) = P_1 f_1(q,p;0) + P_2 f_2(q,p;0) \tag{19}$$

with $f_1 = 0$ when $q,p \notin \Gamma_1^0$ and $f_2 = 0$ when $q,p \notin \Gamma_2^0$ in phase space, with $\Gamma_1^0 \cap \Gamma_2^0 = 0$. Call q_i^0, p_i^0 the mean values of q, p in the distribution f_i and Δq_0, Δp_0 their mean square values, which we assume for simplicity to be the same for f_1 and f_2. Suppose furthermore that the space distance d_0 between Γ_1^0 and Γ_2^0 is $\gg \Delta q_0$. Now we measure q with a resolution Δq_0 and find the particle in S_1^0, the space width of Γ_1^0. We might as well have measured p with a resolution Δp_0, with the result that we would have found the particle in M_1^0, the momentum width of Γ_1^0. In classical mechanics of course both measurements are compatible, but one is sufficient, in this case, to deduce from (19) that at $t=0$ the point in phase space representing the particle's state is in Γ_1^0. Then we have two possible interpretations of this fact:

(a) we can say that even before our measurement at an earlier time t the phase space point of the system was in Γ_1^t (the region which subsequently evolved according to Liouville into Γ_1^0 at $t=0$) because it has followed a trajectory which, starting from a point located within Γ_1^t goes through a point in Γ_1^0. The position of the particle at the earlier time was therefore within a distance $\Delta q_t \ll d_t$ from the mean value q_1^t given by $f_1(q,p;t)$, with Δq_t given by (suppose for simplicity that the particle has propagated freely such that Δp_0 does not vary with time):

$$\Delta q_t = [(\Delta q_0)^2 + (\Delta p_0 t/m)^2]^{1/2} \tag{20}$$

It should be stressed at this point that, while the volume of Γ_1^t is equal to the volume of Γ_1^0, the uncertainty product $\Delta q_t \Delta p_t$ always increases with time (both in the backward and in the forward direction) because of (20).

The probabilities $P_1 P_2$ in (19) represent therefore our ignorance about the previous localization of the particle and not an actual indetermination of its position in space.

(b) we can say that before the measurement there was no phase space point representing the particle's state in $\Gamma_1{}^t$ or in $\Gamma_2{}^t$ and that therefore the state has been localized in $\Gamma_1{}^0$ by the measurement. In this case P_1 and P_2 are intrinsic probabilities of localizing the particle either in $\Gamma_1{}^0$ or in $\Gamma_2{}^0$. There is no trajectory followed by the particle from one point of $\Gamma_1{}^t$ to a given point of $\Gamma_1{}^0$.

In both cases, after the measurement the state is no longer represented by the distribution function $f(q,p)$ but is reduced to $f_1(q,p)$, the new state created by the measurement, which evolves successively according to the Liouville equation. However, in the first case the state $f_1(q,p)$ is the state of a new ensemble in which the states of the individual particles are known only within the corresponding uncertainties, but in the second case there is no difference between the state of the particles and the state of the ensemble. Therefore one immediately recognizes that (a) is the usual interpretation of statistical mechanics in terms of classical dynamics, and (b) is an interpretation which closely resembles the conventional interpretation of quantum mechanics in which the observer has an essential role. In spite of the fact that they both lead to the same observable consequences, our choice is biased in favour of the first one by our belief in the existence of an objective world outside our mind.

Let us now consider the corresponding situation in quantum mechanics. Take a state defined by the wave function

$$\psi = \sqrt{P_1}\,\psi_1 + \sqrt{P_2}\,\exp(i\phi)\,\psi_2 \tag{21}$$

whose Wigner function tends in the classical limit to (19)

$$W(q,p;0) \rightarrow f(q,p;0) \tag{22}$$

The wave functions ψ_1 and ψ_2 have therefore the same mean values and mean square values of q and p as before. Let us assume Δq_0 to be related to Δp_0 by the minimum uncertainty:

$$\Delta q_0 \Delta p_0 \approx \hbar\,/2. \tag{23}$$

Suppose we measure q with the resolution Δq_0 and find the particle within the space region $S_1{}^0$ which is the space support of ψ_1. We might as well have measured p with resolution Δp_0, with the result that we would have found the momentum of the particle in $M_1{}^0$, the momentum support of ψ_1. In both cases we deduce that the state of the particle at t=0 is represented by ψ_1, and evolves subsequently according to the Schrödinger equation. It should be stressed that also in the quantum case the two measurements are compatible, because the two resolutions satisfy the uncertainty relation (23). Both these measurements, therefore, reduce the state (21) but do not change the form of ψ_1. However, according to the conventional interpretation of quantum mechanics, we cannot infer, from this fact that the particle was in $S_1{}^t$ (or $M_1{}^t$) at an earlier time t, because we have to accept that the particle has been located in that region by the act of measurement, and that any statement about its position (or momentum) before the measurement is actually meaningless. Eq.(22), however, forces us to extend this interpretation also to classical statistical mechanics and therefore to adopt interpretation (b), because $S_1{}^t$ ($M_1{}^t$) is the space (momentum) extension of $\Gamma_1{}^t$. *We find therefore an inconsistency if we insist to accept the conventional interpretation (a) for classical statistical mechanics while retaining the conventional interpretation of quantum mechanics.*

5. CONSISTENCY REQUIREMENT BETWEEN CLASSICAL AND QUANTUM INTERPRETATIONS

The standard interpretation of quantum mechanics is therefore incompatible with the usual assumption that newtonian dynamics for individual particles underlies the description of classical statistical ensembles. This suggests that the introduction of the notion of a sort of localization of particles in space should implement the conventional formulation of quantum mechanics. This localization, of course, should always be consistent with the minimum uncertainty allowed by the Heisenberg principle. In other words we believe that one may describe the time evolution of a particle's state in terms of a sort of fuzzy trajectory which is undefined within the region of minimum uncertainty, but is sufficiently localized in phase space to exclude that it may instantaneously jump from one small region to another one very far away.

We are not going to construct explicitly a new theory of this sort. We wish however to examine in more detail whether the possibility exists of modifying the standard interpretation of quantum mechanics in order to save our traditional picture of classical mechanics.

We have dealt up to here with the problem of giving a meaning to the statement that a particle was localized into one or the other of two widely separated regions in space even before an actual measurement of its position has been made. In this case the resolution Δq_0 is given by the width of each wave packet at the time of measurement. Suppose now one localizes a particle in a space region of extension Δq_0 around a value q_0 within a wave packet of larger extension. Does it still make sense to ask the question: where was the particle (again supposed to propagate freely) at an earlier time t?

The answer requires a brief discussion of the analogous classical case. Given a distribution function $f(q,p,0)$ in phase space with mean square values Δq, Δp of q and p, we can reduce our ignorance by measuring both q and p with resolutions Δq_0, Δp_0 such that their product is much smaller than the product $\Delta q \, \Delta p$. Of course in classical mechanics we may chose these resolutions as small as we like (or at least as small as our instruments allow us to do so). Eq.(20) will therefore again give us the uncertainty of the position of the particle at an earlier time t, in terms of the values chosen for these resolutions.

We may now go back to the quantum case described by a wave function ψ whose Wigner function tends to $f(q,p;0)$ in the classical limit. The uncertainties in q and p given by ψ are now such that $\Delta q \, \Delta p \gg h$. Again we may reduce our ignorance by measuring q and p with resolutions $\Delta q_0 \, \Delta p_0$ such that their product is $\ll \Delta q \, \Delta p$ but, of course, we cannot make them as small as we like because of the minimum uncertainty relation (23). These ideal measurements however can be performed in such a way as to minimize the uncertainty in the position of the particle (again supposed to propagate freely) at an earlier time t. We obtain from (20), with the replacement $\Delta p_0 = h / 2 \, \Delta q_0$,

$$\Delta q_t = \min \, [(\Delta q_0)^2 + (h \, t / 2m \Delta q_0)^2]^{1/2} \qquad h / 2 \, \Delta p < \Delta q_0 < \Delta q \qquad (24)$$

where the minimum is taken with respect to Δq_0 and depends on t.

For small times one has $\Delta q_t \approx h / 2m \, \Delta q$; for intermediate times $\Delta q_t \approx (h \, t / 2m)^{1/2}$, and for large times $\Delta q_t \approx h \, t / 2m \, \Delta q$.

This result shows that, even if we cannot precisely localize the particle on a trajectory as in classical mechanics, it is still possible to give an upper limit for the extension of the region where the particle was localized before the measurements. This statement, of course, does not conflict in any way with the physical predictions of quantum mechanics, but leads to the correct newtonian trajectories when the classical limit is performed.

In order to understand fully the meaning of our point of view, we stress again that the effect of a measurement which reduces a wave packet with uncertainty product $\Delta q \Delta p >> h$ into a wave packet with uncertainty $\Delta q_o \Delta p_o \approx h$, is substantially different from a change in the form of a wave packet which maintains the uncertainty equal to its minimum value. It is very important to avoid confusions between the two. The first one is irreversible, because our knowledge changes irreversibly. It implies, exactly as it does in classical mechanics, the measurement of both q and p, the only difference with classical mechanics being that now the resolutions Δq_o and Δp_o must satisfy the minimum uncertainty relation. The state of the individual particle is not reduced: it is only the ensemble state which is reduced. This measurement eliminates the "empty waves" of a superposition because they are not physical: they only represent our ignorance before the measurement.

The second change is reversible, because it corresponds to an actual change of the individual particle's physical state from a wave packet with $\Delta q'_o \Delta p'_o \approx h$ to a wave packet with $\Delta q_o \Delta p_o \approx h$ due to its Schrödinger evolution in presence of a physical interaction. Clearly, there is no reduction in this case, because there is no change in the information we have on the properties of the individual system: what we gain in the definition of q (if $\Delta q_o < \Delta q'_o$), we lose in the definition of p ($\Delta p_o > \Delta p'_o$) and viceversa.

6. THE STATE VECTOR COLLAPSE

We may now explicitly come to the problem of state vector collapse. In the usual formulation (D'Espagnat 1976), a microsystem S, whose state $|\psi\rangle$ is a superposition of eigenstates $|\phi_n\rangle$ of a quantum variable **G** with eigenvalues γ_n, is brought in interaction with a macroscopic measuring instrument M, made of a very large number N of particles whose states $|n\rangle$ are labeled by the eigenvalues γ_n of a macroscopic physical variable Γ, establishing in this way a one-to-one correspondence between the set of γ_n and the set of γ_n. The state W of the total system $S+M$ is therefore, after the interaction

$$\Omega = \sum_n c_n |\phi_n\rangle \, |n\rangle \tag{25}$$

The corresponding density matrix can be written as

$$\rho_\Omega = \Sigma_{nn'} \, c_n c_{n'}^* \mid \phi_n><\phi_{n'} \mid \quad \mid n><n' \mid \tag{26}$$

Suppose now that the Wigner functions $W_n(q,v)$ of the intensive center of mass variables q, v of M constructed with the states $\mid n>$ tend to the set of phase space classical distributions $f_n(q,v)$ which correspond to the disconnected phase space regions in which Γ has the different values γ_n. From the discussion at the end of section 2 (eqs. (16)-(18)) we can say that

$$W_n(q,v) \to f_n(q,v) \qquad N\to\infty \tag{27}$$

and, furthermore, that the contribution of the cross terms with $n\neq n'$ in (26) to $W(q,v)$ have an oscillating behaviour of the form

$$K \cos[(\gamma_n - \gamma_{n'})\alpha] \tag{28}$$

where α is the variable conjugate to Γ and K is a slowly varying function of the variables q,v. These contributions therefore vanish in the limit $N\to\infty$ for the same reasons that made the cross term of the Wigner function of the state (15) vanish in the same limit. The statistical properties of the system $S+M$ described by the density matrix (26) are therefore identical to those described by the reduced density matrix

$$\rho_s(q,v) = \Sigma_n \mid c_n \mid^2 f_n(q,v) \mid \phi_n><\phi_n \mid \tag{29}$$

which is a classical distribution function in the phase space of M and a density matrix for the variables of S. The meaning of (29) is of course that, from a statistical point of view, everything happens as if the wave packet reduction had occurred for the quantum variables of S after the interaction with M (Cini 1983).

The main objection to consider this line of reasoning as a satisfactory solution of the wave packet collapse problem is that eq.(29) represents only the properties of a statistical ensemble of systems $S+M$ but does not represent the actual outcome of each individual act of measurement. In other words it does not explain why, after the interaction with S, the instrument M should objectively be in a given state with a well determined value, say γ_k, of

Γ independently of whether it is observed or not, while the superposition (25), according to the standard interpretation of quantum mechanics, collapses in a well determined state characterized by the value γ_k of Γ only when M, in its turn, is observed.

It is clear that the answer to this objection is provided by our interpretation of quantum mechanics. If our point of view is correct the minimum uncertainty product is completely negligible for the variables q and v of M and all the uncertainty arising from the sum over n in eq.(29) is of statistical nature, namely arises from our ignorance of the actual state of M, which, however, does have a definite value of Γ, independently of our knowledge. The reduction of the mixture (29) to the single term

$$f_k(q,v) \quad |\phi_k><\phi_k| \tag{30}$$

is only a matter of reduction of ignorance, not a physical phenomenon in which M is involved.

To illustrate the situation with a practical example let us consider a Stern Gerlach device in which a beam of particles initially located in a given space region is split, by means of an inhomogeneous magnetic field interacting with the magnetic moment of each particle, in two beams whose separation increases with time. We study the time evolution of their lateral widths which at t=0 coincide. The initial wave packet

$$\Psi(x,0) = \psi(x,0) \, [c_+u_+\exp(ip_ox/h) + c_-u_-\exp(-ip_ox/h)] \tag{31}$$

(where $\psi(x,0)$ is given by (10) and $u_\pm$ are the spin up and down eigenfunctions) gives rise, after a time t, to a mixed density matrix which is the sum of two widely separated parts:

$$\rho_s(x,p;t) = P_+W_+(x,p;t) \, |u_+><u_+| + P_-W_-(x,p;t) \, |u_-><u_-| \qquad P_\pm = |c_\pm|^2 \tag{32}$$

where $W_\pm(x,p;t)$ are given by eq.(8) with x_o replaced by $x_\pm = x_o \pm p_ot/m$. The interference terms are easily found, as already discussed at length, to be vanishingly small because the overlap of the two gaussians centered at $x_\pm$ is practically zero. The mean square values of x, p are given by (we take the trace on the spin variables):

$$(\Delta x)^2 = (\Delta x_o)^2 + (\Delta p_o)^2 t^2 / m^2 + 4 P_+ P_- p_o^2 t^2 / m^2 \tag{33}$$

$$(\Delta p)^2 = (\Delta p_o)^2 + 4 P_+ P_- p_o^2 \tag{34}$$

Now, if the lateral width of the individual beams is much smaller than their distance (namely if $p_o >> \Delta p_o$, and $t >> h \, m/p_o \Delta p_o$) the uncertainty product reduces to the classical expression

$$(\Delta x \, \Delta p)_{cl} = 4 \, P_+ P_- \, p_o^2 t / m \tag{35}$$

which represents the effect of the uncertainty $\pm p_o$ in the momentum of a particle on the spread of its position's uncertainty. Here again, if we find the particle in the region occupied by the beam with momentum $+p_o$, we have to conclude that it was in that beam even before we made the measurement. The probabilities $P_\pm$ represent therefore our ignorance and not an intrinsic delocalization of the particle.

The case of the Stern Gerlach device shows therefore that our interpretation and the conventional one have very different implications. For us the particle is already in one of the two beams before its detection by a counter which may have been placed on its way. The counter is discharged because the particle is already in the beam which impinges on it. We stress that this does not imply that coherence has been destroyed once for all. In fact, if the two beams are superimposed again, the occupied phase space is not anymore the union of two classically separated regions, and therefore the typical quantum interference occurs again. In our picture the reduction of the wave function is simply a consequence of the additional information acquired on the state of the particle which allows us to change our description of it, and no problem arises.

For the conventional interpretation it is the other way round: the counter's discharge localizes the particle in the region where it occurs, inducing an abrupt change in the physical entity represented by the wave function. In this case, as already discussed at length, one has to explain many puzzling features of this sudden and irreversible change of the particle's properties.

Obviously, once the reduction of the ensemble's state due to a reduction of ignorance is accepted as a feature of quantum mechanics which it has in common with classical statistical mechanics, the absence of reduction of the individual states will not lead to a nightmarish multiplication of worlds (Everett 1957), because the reduction of ignorance is sufficient to eliminate the proliferation (which, by definition, implies a tremendous increase with time of the uncertainty product) of branches of a composite system's wave function. At the same time this absence of reduction is sufficient to eliminate the extremely "counterintuitive mutual involvement of physical and mental phenomena" (Shimony 1963) invoked explicitly by von Neumann but implicitly accepted by all theories of measurement which adopt the wave function collapse postulate as a physical irreversible phenomenon which cannot be reduced to the Schrödinger time evolution.

The interpretation of quantum mechanics proposed here has, on the other side, some similarities with the so called "environmental" theories (Primas 1981) (Zurek 1983) (Joos and Zeh 1985) which attribute to a quantum object the capability of acquiring classical properties as a consequence of its interaction with the environment. They have in common with our approach in a broad sense the idea that quantum objects may acquire, in appropriate conditions, localization properties which are preexistent to their measurement.

There is however an essential difference. For these theories the interaction between instrument and environment eliminates the quantum correlations between the object and the instrument without introducing further correlations between the instrument and the environment, because the state of the latter is never detected. For us the quantum correlations between object and instrument are already washed out by the classical behaviour of the instrument for values of the macroscopic variables which do not show quantum interference effects. Since there are different versions of the "environmental" point of view a detailed comparison would be exceedingly lengthy. However we will attempt to extract their common feature.

In one version (Zurek 1983) it is the instrument M which interacts with the environment E. The initial state Ω of eq.(34) is therefore transformed into a state Ψ of the total system $S+M+E$. By taking the trace over the environment's variables which are not observed one obtains a reduced density matrix

$$\rho^{\circ}{}_{S+M} = \text{Tr}_E (\Psi\Psi^{\dagger}) \tag{36}$$

of the system $S+M$ in which the off diagonal elements of the density matrix ρ_{Ω} constructed with the state Ω are absent. The off-diagonal elements of ρ_{Ω} are however already vanishingly small for all the macroscopic variables of M and there seems no need of further eliminating them.

In another version (Joos and Zeh 1985) it is the system S which is correlated with the environment. In this case the initial state $S+E$

$$\Phi = \sum_\alpha a_\alpha |\varphi_\alpha\rangle \, |E_\alpha\rangle \qquad\qquad (37)$$

becomes, after the interaction between S and M

$$\Psi = \sum_{\alpha n} a_\alpha P_n |\varphi_\alpha\rangle \, |E_{\alpha n}\rangle \, |n\rangle \qquad\qquad (38)$$

where the environment states will be mutually orthogonal in each index. Again, taking the trace over E one obtains a reduced density matrix of the type (36).

In any case the main problem is still there, namely, to justify why we should interpret the density matrix (36) as a classical statistical ensemble, in which each individual system of the ensemble is in a definite, albeit unknown, state, rather than the density matrix of a quantum statistical mixture, in which each individual system is projected into a definite state only when it is observed. The difficulties that still remain in this approach as a consequence of this ambiguity are particularly evident in the Stern Gerlach example discussed above.

In this case one has actually, three variables of different nature. The first variable is the particle's spin, an essentially quantum variable. The second one is the particle's position, which is a quantum variable which may acquire macroscopic values. The third is the counter's charge which is a typical macroscopic variable. The spin, obviously, cannot be reduced to a classical variable by the interaction with the environment. Therefore, it is either the particle's position or the counter's charge which acquires classical properties.

In the first case, however, the environment which induces the collapse of the superposition into one or the other beam when the beams are far apart, is unable to transform back the mixture into a superposition when the beams are brought to overlap again. Since we know that if this happens they do interfere again, it seems difficult to ascribe to the environment the reversible space localization of the two beams that we have envisaged in order to reconcile QM with its classical limit.

In the second case the collapse of the superposition between the two macroscopically different states of the counter (neutral and discharged) may well be ascribed to the environment. In this case, however, one is again forced to assume, as in the standard interpretation, that the particle jumps from one beam to another, even when they are far apart, according to whether the counter is triggered or not by the interaction with the environment. This is indeed a counterintuitive conclusion that we prefer to avoid, if possible.

We think therefore that the elimination of the contradiction between the standard interpretation of quantum mechanics and the accepted interpretation of classical statistical mechanics yields also the solution of the old problem of the measurement induced state vector collapse.

References

Cini M 1983 Nuovo Cim. **73**B 27

D'Espagnat B 1976 Conceptual foundations of Quantum Mechanics (W.A.Benjamin New York)

Einstein A 1909 Phys. Zeitschr. **10** 185

Everett H 1957 Rev.Mod.Phys. **29** 454

Joos E and Zeh H D 1985 Z.Phys.B **59** 223

Primas H 1981 Chemistry, Quantum Mechanics and Reductionism (Berlin: Springer)

Shimony A 1963 Am.J.Phys. **31** 755

Wigner E P 1932 Phys.Rev. **40** 479

Zurek W H 1983 Phys.Rev.D **26** 1862

Consecutive Quantum Measurements

Asher Peres

1. INTRODUCTION AND SUMMARY

In some elementary textbooks of quantum mechanics, the quantum wave function ψ is treated as if it were a physical object, undergoing a "collapse" whenever a measurement is performed (Bohm 1951). This collapse, which is a *nonlinear* transformation, is obviously incompatible with Schrödinger's equation and leads to numerous paradoxes (Schrödinger 1935).

This interpretation of the theory is apparently based on the naive belief that the time evolution of ψ represents what is actually happening in the real world. In fact, there is no experimental evidence whatsoever that every physical system has at every instant a unique, well-defined state ψ (or, possibly, a density matrix ρ) and that the evolution of $\psi(t)$, or of $\rho(t)$, represents an actual physical process (Peres 1988). Quantum theory, correctly interpreted, makes no such claim. It is only a mathematical formalism, allowing us to compute *probabilities* for the occurrence of *macroscopic* events of a specified kind, following a specified preparation (Stapp 1972). The wave function ψ (or the density matrix ρ) represents only the *statistical information* available to the observer (Peres 1984, 1986).

The purpose of this article is to show how a sequence of quantum measurements can be described without any mention of the hypothetical "collapse." To have a concrete example, we consider the detection of a weak classical signal: The quantum system—the *detector*—is driven by an unknown time-dependent force. This may be, for example, a weak electromagnetic or gravitational wave. These weak signals can indeed be treated classically, because they contain enormous numbers of photons or gravitons (Caves *et al* 1980). The problem is to decode the information stored in the quantum system and to convert it into a new classical signal—the reading of a *meter*. This is the classic "quantum measurement" problem. In general, quantum measurements lead to an entropy increase (von Neumann 1955) and therefore to a degradation of information. Thus, in summary, we examine the amount of distortion caused by the presence of a quantum interface between two classical signals. It will be shown that the resolving power of the meter must be matched to the spectral properties of the quantum system. While a poor resolution obviously gives inaccurate results, a resolution that is too sharp is also undesirable, because in that case the meter overwhelms the detector and yields results having little relationship to the original signal.

Section 2 of this paper discusses the *final* link of the amplification chain: the conversion of information encoded in a quantum state into the reading of a classical meter. The treatment is strictly quantum-mechanical. In particular, there never is any "collapse" of a wave function. As explained above, the meaning of the wave function is that of a mere mathematical tool, allowing us to compute *probabilities* for the occurrence of specified macroscopic events. It is shown that the rule saying "the observable values of a dynamical variable are the eigenvalues of the corresponding operator" is valid only in the limit of idealized meters.

Consecutive measurements are discussed in Section 3. There must be as many independent meters as there are data to be taken. Naturally, all these "meters" can be incorporated into a single instrument (Dicke 1989). In the limiting case of very precise measurements performed repeatedly at very short time intervals, the meters *lock* at one of the eigenvalues. This is the so-called "quantum Zeno paradox" (which has nothing paradoxical: the meters literally overwhelm the quantum system).

The general results obtained in Section 3 are illustrated in Section 4 for a model where the unknown signal is a variable torque causing the precession of a rotor (a particle of known spin). First, we consider the simple case of a spin $\frac{1}{2}$ system, for which every detail of the calculation can be followed explicitly. As expected, a spin $\frac{1}{2}$ particle cannot be a good detector: its Hilbert space is too small. We then consider the case of a particle having a large spin. It is shown that arbitrarily weak torques can be observed without appreciably disturbing the precession of the detector, provided that the meters have a resolution suitably matched to the spin spectrum.

Throughout this paper $\hbar = 1$ and, moreover, the unit of length is chosen in such a way that the meter's scale directly gives the eigenvalues of the measured operator.

2. QUANTUM MEASUREMENTS

A "measurement" is a process which generates a *correlation* between a property of the measured object and a property of the meter. As a simple example, consider a particle of spin $\frac{1}{2}$ whose initial state is described by a spinor $\begin{pmatrix} \alpha \\ \beta \end{pmatrix}$. Suppose for simplicity that there are no other degrees of freedom and no forces acting on that particle. The problem is to measure σ_z, a dynamical variable purported to have observable values ± 1.

The "meter" which performs this measurement is idealized as being another particle, with position q, momentum p, and mass M. Its free Hamiltonian is $p^2/2M$. Its interaction with the spin $\frac{1}{2}$ particle is described by

$$H_{\text{int}} = g(t)\,\sigma_z\,p, \tag{1}$$

where $g(t)$ is an externally controlled function of time[1] with narrow support near $t = 0$, and such that $\int g(t)\,dt = 1$, in appropriate units.

Both σ_z and p are constants of the motion. The Heisenberg equation of motion for q is

$$\dot{q} = i[H, q] = g(t)\,\sigma_z + (p/M). \tag{2}$$

The last term of (2) can be neglected during the brief interaction. The solution of (2) thus is

$$q_f = q_i + \sigma_z. \tag{3}$$

This result is better visualized in the Schrödinger picture. Let $\phi(q)$ denote the initial wave function of the meter. Assume that this is a function with a sharp maximum at $q = 0$ (this means that, before the measurement, it is most probable to find the meter close to $q = 0$). Moreover assume that M, the mass of the meter, is so large that $\phi(q)$ will not appreciably spread during the experiment. In other words, it is legitimate to take $H \equiv H_{\text{int}}$. The time evolution generated by this Hamiltonian is

$$\begin{pmatrix} \alpha \\ \beta \end{pmatrix} \phi(q) \;\rightarrow\; \begin{pmatrix} \alpha \\ 0 \end{pmatrix} \phi(q-1) + \begin{pmatrix} 0 \\ \beta \end{pmatrix} \phi(q+1). \tag{4}$$

This process is illustrated in Fig. 1. The meaning of the right hand side of (4) is the following: There is a probability amplitude $\alpha\phi(q-1)$ to find the meter near $q = 1$, and the particle with spin up; and a probability amplitude $\beta\phi(q+1)$ to find the meter near $q = -1$, and the particle with spin down. I emphasize that quantum

---

[1] In a real-life Stern-Gerlach experiment, the time dependence of $g(t)$ would be due to the passage of the particle through a localized magnetic field. This can be described by a time-*independent* Hamiltonian, involving an additional degree of freedom (the position of the spin $\frac{1}{2}$ particle). For details, see Peres (1980a).

mechanics allows to compute only probabilities of events. It does not describe the events themselves (Komar 1962).

What we call "the observed value of σ_z" is given by the final position of the meter. The wave function ϕ gives the probabilities of observing the meter at its possible final positions. Ideally, $\phi(q \mp 1)$ should have zero width and the result of the measurement should be ± 1, *i. e.*, one of the eigenvalues of σ_z. However, the actually observed value of σ_z may differ from the ideal result by a quantity of order Δq, the width of the meter wavepacket. This discrepancy is not a trivial "technical difficulty," but a matter of principle. It will be seen in the following sections that in some cases it may be necessary to have Δq larger than the separation of consecutive eigenvalues.

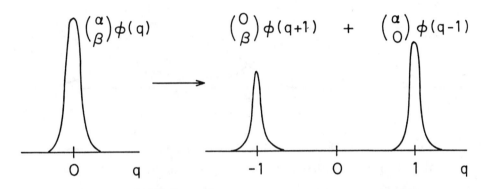

FIG. 1. The evolution of the wave function due to a measurement process. In this drawing and the following one, $\alpha = 0.8$ and $\beta = 0.6$ (the vertical scales are in arbitrary units).

Although there is no advantage to such a situation in the present problem (which involves a single measurement of σ_z) let us examine the consequences of having $\Delta q > 2$. The situation is represented in Fig. 2. The probability of observing the meter between q_0 and $q_0 + dq_0$ is

$$P \, dq_0 = [\, |\alpha|^2 |\phi(q_0 - 1)|^2 + |\beta|^2 |\phi(q_0 + 1)|^2 \,] \, dq_0, \qquad (5)$$

where both terms may have contributions of the same order of magnitude.

For future reference, it is convenient to rewrite the preceding equations in terms of density matrices. Let λ_m be the eigenvalues of the operator A being measured (in the simple case considered above, A was σ_z and we had $\lambda_m = \pm 1$). Let ρ_{mn} be the density matrix of the quantum system, in a representation where A is diagonal. Let

$\Phi(q', q'') = \phi(q')\phi^*(q'')$ be the initial density matrix of the meter. The combined density matrix thus is

$$\rho_{mn}(q', q'') = \rho_{mn}\, \Phi(q', q''). \tag{6}$$

The interaction Hamiltonian (1) becomes $g(t)Ap$ and, instead of (4), we now have

$$\rho_{mn}(q', q'') \rightarrow \rho_{mn}(q' - \lambda_m, q'' - \lambda_n). \tag{7}$$

This expression contains all the information about the combined state of the quantum system and the meter used to observe it.

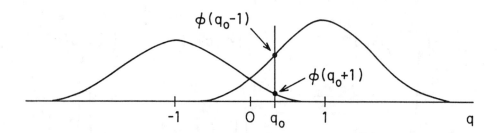

FIG. 2. This is the same as the right hand side of Fig. 1, except that the initial location of the meter is uncertain by more than the separation of the eigenvalues ± 1, so that the measurement is "fuzzy."

At this point, if we are no longer interested in the quantum system, we may trace out the indices referring to it. Then, the average value of any observable function $f(q)$—after completion of the interaction between the meter and the quantum system—is

$$\langle f(q) \rangle = \sum_j w_j \int \Phi(q - \lambda_j, q - \lambda_j)\, f(q)\, dq, \tag{8}$$

where $w_j = \rho_{jj}$ is the probability of occurence of the jth eigenstate of A, just before (or after) the interaction. For example, in the simple case where the meter is prepared in a symmetric state, $\phi(q) = \phi(-q)$, we have

$$\langle q \rangle = \sum_j w_j \lambda_j, \tag{9}$$

and

$$\langle q^2 \rangle = \sum_j w_j \left(\int q^2 |\phi(q)|^2 dq + \lambda_j^2 \right). \tag{10}$$

Therefore

$$\langle q^2 \rangle - \langle q \rangle^2 = (\Delta A)^2 + (\Delta q)^2, \tag{11}$$

where

$$(\Delta A)^2 = \sum_j w_j \lambda_j^2 - \left(\sum_j w_j \lambda_j \right)^2, \tag{12}$$

and

$$(\Delta q)^2 = \int q^2 \, |\phi(q)|^2 \, dq, \tag{13}$$

are the variances associated with the quantum system and the meter, respectively.

Conversely, if we are interested in the quantum system (for later use) but not in the meter itself, the new density matrix ρ'_{mn} of the quantum system is obtained by tracing out q' and q'':

$$\rho'_{mn} = \int \rho_{mn}(q - \lambda_m, q - \lambda_n) \, dq = \rho_{mn} S_{mn}, \tag{14}$$

where S_{mn} is the *coherence matrix*:

$$S_{mn} = \int \phi(q - \lambda_m) \, \phi^*(q - \lambda_n) \, dq = \langle e^{ip(\lambda_m - \lambda_n)} \rangle. \tag{15}$$

For example, if

$$\phi(q) = (2\pi\sigma)^{-1/4} \, e^{-q^2/4\sigma^2}, \tag{16}$$

we have $\Delta q = \sigma$ and

$$S_{mn} = \exp[-(\lambda_m - \lambda_n)^2/8\sigma^2]. \tag{17}$$

Obviously, $S_{mm} = 1$ (for any ϕ) so that the diagonal matrix elements of ρ are not affected. On the other hand, the off-diagonal elements of ρ are depressed by a factor S_{mn}, and may even be reduced to zero if the displaced wave functions are mutually orthogonal, as in Fig. 1. However, if $\Delta q \gtrsim |\lambda_n - \lambda_{n-1}|$, as in Fig. 2, the off-diagonal elements of ρ are not completely suppressed. (In particular, if some eigenvalues are degenerate, the submatrix of ρ_{mn} corresponding to these eigenvalues is not affected at all.)

In the following sections, we shall consider a sequence of consecutive measurements. In particular, it will be seen that a low resolution such as the one illustrated in Fig. 2 may sometimes be advantageous when we want to monitor the time evolution of a dynamical variable. Some authors (Kraus 1983, Braunstein and Caves 1988) discuss this problem in the formalism of *effects* and *operations*. The present paper uses the standard Hamiltonian formalism which is, in my opinion, simpler and clearer.

3.   CONSECUTIVE MEASUREMENTS

The detection and analysis of time-dependent signals necessitates numerous measurements, distributed in time (Caves 1986 and 1987, Lamb 1987) so as to get a *sequence of numbers*, corresponding to times t_1, t_2, ... and so on. There can be no continuous measurement, because a "measurement" was defined as a *brief and intense interaction* between the meter and the measured system, as shown for example in Eq. (1). In particular, the support of $g(t)$ must be smaller than the difference $t_k - t_{k-1}$, so that the measurements do not overlap.

It is possible for sure to consider measurements of finite duration, where the function $g(t)$ is spread over an appreciable time, but such a measurement yields only a *single number* corresponding, in the best case, to a time-average of the observed variable (Peres and Wootters 1985). One can even consider a passive detector, such as a Geiger counter waiting for the decay of a nucleus, but this situation does not fit at all with our definition of a measurement. This setup is best described as a single metastable system with several decay channels (Peres 1987).

Quantum theory by itself does not impose any fundamental limitations to monitoring arbitrarily weak signals within arbitrarily short time intervals. Limitations arise solely because we want to use—or are forced to use—some particular detectors. For example, gravitational radiation couples very weakly to matter, and detectors must be very massive, and expensive, antennas. This means, effectively, that it is impracticable to have a large number of identical detectors from which quantum averages are to be obtained. We must extract as much information as possible from *each* measurement.

Two strategies are possible. We may prepare the detector in a known state, which is an eigenstate of its free Hamiltonian (that is, the detector's Hamiltonian when no signal is present). We wait some preassigned time and then measure an operator which commutes with the free Hamiltonian. If the measurement is sharp, as in Fig. 1, the detector is left in a known eigenstate and the process can be repeated. This stategy is conceptually simple, but very inefficient: Each resetting of the detector destroys latent information, namely the relative amplitudes and phases of the various components of the detector's wave function just prior to the measurement (one component is selected by the measurement, the other ones are lost forever). In particular, if the time t elapsed between propagation and observation is smaller than $(\pi/2\Delta H)$, where

$$\Delta H = [\langle H^2 \rangle - \langle H \rangle^2]^{1/2}, \tag{18}$$

is the energy uncertainty of the detector (including its coupling to the signal), there is a probability larger than $\cos^2(t\,\Delta H)$ that the *initial* eigenstate will again be observed

(Fleming 1973, Peres 1980b). Very weak signals therefore necessitate to wait a long time between consecutive measurements. How long is unpredictable, in the absence of extraneous information. Finally, if and when enough nontrivial data have been obtained, we have to reconstruct from these data the time dependence of the signal. For example, this time dependence can be represented by a trial function including some unknown parameters, and the latter can then be fitted to the experimental data.

A more efficient strategy is to let cumulative effects of the signal develop while the measurements are being performed. We shall see that this is possible if the measurements are "fuzzy," as illustrated in Fig. 2. When Δq (of the meter) is much larger than the separation of consecutive eigenvalues (of the detector), the relative amplitudes and phases of the corresponding wave function components are not completely lost. The dynamical evolution of the detector can thus proceed, although it cannot be the same as in the absence of measurements. The reason is that if we want to measure a time-dependent variable $A(t)$, and if $[A(t), A(t')] \neq 0$, a measurement of $A(t)$ "disturbs the value of $A(t')$." Stated more precisely: Given an ensemble of identically prepared and identically measured systems, the histogram of observed values of $A(t')$ depends on whether or not there is a prior measurement of $A(t)$.

Although the evolution of a quantum detector is inevitably modified by continually measuring it, we would like the meters' readings to remain reasonably reliable, even after numerous observations. Ideally, we would like the quantum evolution to mimic the classical one, and the output signal to be an amplified replica of the input signal. The signal distortion (and concomitant loss of information) should be minimized. The purpose of this paper is to investigate how closely these aims can be achieved.

Let us consider a sequence of measurements performed at times $t_1, \ldots, t_N$, by means of meters with coordinates $q_1, \ldots, q_N$, respectively. The initial density matrix of the detector and the meters is a generalization of (6):

$$\rho_{mn}(q_1', q_1'', \ldots, q_N', q_N'') = \rho_{mn} \prod_j \Phi_j(q_j', q_j''). \tag{19}$$

The Hamiltonian of the combined system is

$$H = H_0(t) + A \sum_j g(t - t_j) p_j, \tag{20}$$

with the same notations as in Eq. (1). Here H_0 involves only the dynamical variables of the detector and in particular H_0 has a known functional dependence on the unknown signal. The masses of the meters are assumed so large that we can neglect their contributions $p_j^2/2M_j$ to H_0. In other words, we can ignore the spontaneous spreading of each meter's wave packet.

In the interval between measurements, H_0 generates the unitary evolution $\rho \to U\rho U^\dagger$. On the other hand, circa each $t = t_j$, there is an evolution similar to Eq. (7), namely

$$\rho_{mn}(\dots, q_j', q_j'', \dots) \to \rho_{mn}(\dots, q_j' - \lambda_m, q_j'' - \lambda_n, \dots). \tag{21}$$

Note that the new density matrix entangles in a nontrivial way the discrete indices mn of the detector and the coordinates of the jth meter.

Equation (21) contains all the information about the state of the detector and the various meters. We can then ask a variety of questions, such as those at the end of the preceding Section. For example, if we are interested only in the detector, not in the meters that have already interacted with it, the net result of a measurement is given by Eq. (14):

$$\rho_{mn} \to \rho_{mn}' = \rho_{mn} S_{mn}. \tag{22}$$

After that, the following meter, if observed, will give results similar to (9) and (10), with

$$w_j = \sum_{mn} U_{jm}\, \rho_{mn}'\, U_{jn}^*, \tag{23}$$

where U is the unitary matrix representing the free evolution of the detector since the preceding measurement, which left the detector in state ρ_{mn}'.

We may also be interested in comparing the readings of different meters. For example, if there are N consecutive measurements, let us predict the expected $\langle (q_1 - q_N)^2 \rangle$ (regardless of the results obtained at $t_2, \dots, t_{N-1}$). As before, let ρ_{mn} be the density matrix of the detector just before the first measurement. The latter causes

$$\rho_{mn} \Phi_1(q_1', q_1'') \to \rho_{mn} \Phi_1(q_1' - \lambda_m, q_1'' - \lambda_n) \equiv \rho_{mn}^{(1)}(q_1', q_1''), \tag{24}$$

(the q_j referring to subsequent measurements are omitted, for brevity). Between the first and second measurements, there is a unitary evolution $\rho^{(1)} \to \rho^{(2)} = U^{(1)}\rho^{(1)}U^{(1)\dagger}$ so that, just before t_2, we have

$$\rho_{rs}^{(2)}(q_1', q_1'') = \sum_{mn} U_{rm}^{(1)} U_{sn}^{(1)*}\, \rho_{mn}^{(1)}(q_1', q_1''). \tag{25}$$

Then, there is a second measurement whose result is ignored (q_2' and q_2'' are traced out) because, as already stated, our present problem is to predict the expected value of $\langle (q_1 - q_N)^2 \rangle$, regardless of the results obtained at intermediate times. According to (14) this leads to a reduction of the off-diagonal elements $\rho_{rs}^{(2)} \to \rho_{rs}^{(2)} S_{rs}$. Consider in particular the case where the measurements are sharp (as in Fig. 1) so that $S_{rs} = \delta_{rs}$.

We thus have, immediately after t_2, a density matrix containing only the diagonal elements $\rho_{rr}^{(2)}$ of (25).

Between the second and the third measurement, there is another unitary evolution $\rho^{(2)} \to \rho^{(3)} = U^{(2)} \rho^{(2)} U^{(2)\dagger}$, with result

$$\rho_{st}^{(3)}(q_1', q_1'') = \sum_r U_{sr}^{(2)} U_{tr}^{(2)*} \rho_{rr}^{(2)}(q_1', q_1''). \tag{26}$$

Again, the third measurement is sharp and only the diagonal elements survive. These are

$$\rho_{ss}^{(3)}(q_1', q_1'') = \sum_r V_{sr}^{(2)} \rho_{rr}^{(2)}(q_1', q_1''), \tag{27}$$

where

$$V_{sr}^{(2)} = |U_{sr}^{(2)}|^2. \tag{28}$$

The rule to continue is obvious. After the last unregistered measurement at t_{N-1}, we have a diagonal density matrix with elements

$$\rho_{xx}^{(N-1)}(q_1', q_1'') = \sum_{wv..} \sum_{..sr} V_{xw}^{(N-2)} V_{wv}^{(N-3)} \cdots V_{ts}^{(3)} V_{sr}^{(2)} \rho_{rr}^{(2)}(q_1', q_1''). \tag{29}$$

Then, finally, the Nth measurement gives

$$\rho_{zy}^{(N)}(q_1', q_1''; q_N', q_N'') = \sum_x U_{zx}^{(N-1)} U_{yx}^{(N-1)*} \rho_{xx}^{(N-1)}(q_1', q_1'') \, \Phi_N(q_N' - \lambda_z, q_N'' - \lambda_y). \tag{30}$$

Recall that we are interested in $\langle (q_1 - q_N)^2 \rangle$, for which we need only the reduced density matrix involving q_1 and q_N, with the detector's indices traced out. This is

$$\rho(q_1', q_1''; q_N', q_N'') = \sum_{zr} W_{zr} \, \rho_{rr}^{(2)}(q_1', q_1'') \, \Phi_N(q_N' - \lambda_z, q_N'' - \lambda_z), \tag{31}$$

where

$$W_{zr} = \sum_{x \dots s} V_{zx}^{(N-1)} V_{xw}^{(N-2)} \cdots V_{ts}^{(3)} V_{sr}^{(2)}, \tag{32}$$

and

$$\rho_{rr}^{(2)}(q_1', q_1'') = \sum_{mn} U_{rm}^{(1)} U_{rn}^{(1)*} \rho_{mn} \, \Phi_1(q_1' - \lambda_m, q_1'' - \lambda_n). \tag{33}$$

Here, we may be tempted to increase N and to make the time intervals very short, so as to have a quasi-continuous monitoring. However, as we shall presently see, this would lead to a complete *loss* of information. Consider in particular the case where the measurements are equally spaced, at intervals $\tau = (t_N - t_1)/(N - 1)$, and let $N \to \infty$ while $T = t_N - t_1$ is kept fixed. Let

$$h = \int_{t_n}^{t_n+\tau} H(t) \, dt. \tag{34}$$

We then have, in the time τ,

$$U = e^{-ih} = I - ih - \tfrac{1}{2}h^2 + \dots, \tag{35}$$

whence

$$V_{sr} \equiv |U_{sr}|^2 = \delta_{sr} - \delta_{sr} \sum_j |h_{sj}|^2 + |h_{sr}|^2 + O(h^4). \tag{36}$$

It follows that each V in (32) differs from the unit matrix by terms of order $(TH/N)^2$. Since there are $N - 2$ such terms in (32), $W_{zr} = \delta_{zr} + O(T^2H^2/N)$. Thus, in the limit $N \to \infty$, we can replace all the U and V by unit matrices and we obtain

$$\rho(q_1', q_1''; q_N', q_N'') = \sum_j w_j \, \Phi_1(q_1' - \lambda_j, q_1'' - \lambda_j) \, \Phi_N(q_N' - \lambda_j, q_N'' - \lambda_j), \tag{37}$$

where $w_j = \rho_{jj}$, as usual. We then have, by virtue of (9) and (10),

$$\langle (q_1 - q_N)^2 \rangle = (\Delta q_1)^2 + (\Delta q_N)^2. \tag{38}$$

The dynamical evolution of the detector has been "frozen" by its continual interaction with the meters. This is the well known quantum Zeno effect (Misra and Sudarshan 1977, Chiu *et al* 1977, Peres 1980c, Singh and Whitaker 1982, Zurek 1984).

    This effect was proved here without invoking the unnecessary "collapse" postulate. It has nothing paradoxical, notwithstanding its name "Zeno paradox." What happens simply is that the quantum system is overwhelmed by the meters which continually interact with it. Note that the derivation essentially depends on the assumption $S_{mn} = \delta_{mn}$ or, in other words, $\Delta q \ll |\lambda_j - \lambda_{j-1}|$. Meters with a coarser resolution do not completely block the detector's motion. Indeed, it was shown by Caves and Milburn (1988) that τ can be made arbitrarily small, provided that σ increases as τ^{-1}.

4. EXAMPLE: DETECTION OF A WEAK TORQUE

As a concrete example, assume that a random signal $\omega(t)$ can be used as a torque acting on a rotor with angular momentum $\mathbf{J}$. The Hamiltonian of the rotor is

$$H = H_0 + \omega(t) \, J_y. \tag{39}$$

If the rotor is spherically symmetric, $H_0 = \mathbf{J}^2/2I$ is a constant of the motion, which can be ignored. The equations of motion (in classical or quantum theory) are

$$\dot{J}_x = J_z, \qquad \dot{J}_y = 0, \qquad \dot{J}_z = -J_x. \tag{40}$$

Their solution is

$$J_x = J_{x0} \cos\tau + J_{z0} \sin\tau, \tag{41a}$$

$$J_y = J_{y0}, \tag{41b}$$

$$J_z = -J_{z0} \sin\tau + J_{x0} \cos\tau. \tag{41c}$$

where

$$\tau = \int_0^t \omega(t)\, dt. \tag{42}$$

In classical physics, the initial values J_{k0} are known and continual measurements of J_x and J_z give $\tau = \tau(t)$:

$$\tau = \tan^{-1}(J_x/J_z) - \tan^{-1}(J_{x0}/J_{z0}), \tag{43}$$

whence we can obtain $\omega(t)$. Unfortunately, this method is not readily applicable to quantum systems, because J_x and J_z do not commute and therefore (43) is not valid. We shall now see in detail what can be done when the "rotor" is a particle of spin $\frac{1}{2}$. Thereafter, we shall consider a particle of large spin j.

The state of a spin $\frac{1}{2}$ particle can be described by a density matrix

$$\rho = \tfrac{1}{2}(I + \mathbf{m}\cdot\boldsymbol{\sigma}), \tag{44}$$

where $\mathbf{m} = 2\langle\mathbf{J}\rangle$. Since the equations of motion (41) are linear, they are satisfied by $\mathbf{m}$ as well as by $\mathbf{J}$. In this example, we shall assume that initially $\mathbf{m} = (0,0,1)$. Therefore $\mathbf{m}$ will remain in the xz plane. The coherence matrix S_{mn} has diagonal elements 1, and off-diagonal elements $S_{12} = S_{21} = S < 1$ (assumed real, for simplicity). Thus, in the present notation, the density matrix reduction (14) simply is

$$m'_x = Sm_x, \qquad \text{and} \qquad m'_z = m_z. \tag{45}$$

The combined effect of a rotation and a reduction is the mapping

$$\begin{pmatrix} m_x \\ m_z \end{pmatrix} \rightarrow \begin{pmatrix} m'_x \\ m'_z \end{pmatrix} = \Omega \begin{pmatrix} m_x \\ m_z \end{pmatrix}, \tag{46}$$

where

$$\Omega = \begin{pmatrix} S\cos\theta & S\sin\theta \\ -\sin\theta & \cos\theta \end{pmatrix}. \tag{47}$$

Here, $\theta = \int \omega(t)dt$ during the time interval since the preceding measurement. Recall that, for a Gaussian shaped $\phi(q)$, we have $S = \exp[-1/8(\Delta q)^2]$. The extreme cases are $\Delta q \gg 1$, giving $S \simeq 1$ and therefore an unperturbed rotation of the vector **m** (but no true measurement, of course); and $\Delta q \ll 1$ (whence $S \simeq 0$) corresponding to a sharp measurement.

Suppose that we repeatedly measure J_z. Initially, we have $\langle J_z \rangle = \frac{1}{2}$. Thereafter, all we can obtain is a sequence of $+\frac{1}{2}$ and $-\frac{1}{2}$ from which we have to reconstruct the function $\tau(t)$: Obviously, a spin $\frac{1}{2}$ particle is not a good torque detector—its Hilbert space is too small. On the other hand, these particles come cheap (that is, for gedankenexperiments) so that we can afford to use a large number N of identical detectors. It therefore makes sense to compute the expected average $m_z = \langle J_z \rangle$, under various scenarios.

The following figures illustrate the behavior of $m_z(t)$, in the simple case $\omega = 1$ (so that $\tau = t$) from $t = 0$ to $t = 5\pi/2$. The dotted line is the undisturbed $m_z = \cos t$ evolution, corresponding to $\Delta q \gg 1$ (we would of course need $N \gg (\Delta q)^2$ to actually

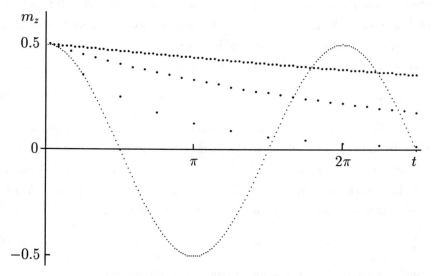

FIG. 3. Average $\langle J_z \rangle$ of a spin $\frac{1}{2}$ particle precessing around the y axis with constant $\omega = 1$. Initially, $\langle J_z \rangle \equiv m_z = 0.5$. The dotted line corresponds to the undisturbed evolution (that is, each point represents the value of m_z which would be obtained if there were no measurement before that time). The circles represent consecutive values of m_z that are observed if *sharp* measurements are performed at intervals (from top to bottom) $5\pi/180$, $5\pi/60$, and $5\pi/20$.

observe it, as an average over N data). This ideal result is compared, in Fig. 3, to the case $S = 0$ (sharp measurements) for 10, 30, and 90 equally spaced samplings. Obviously, the more frequent the measurements, the less m_z moves (this is the Zeno effect). In every case, m_z decays exponentially—it does not oscillate as the unperturbed m_z. This can be verified by computing the eigenvalues of Ω in (47). The latter are given by the secular equation $\lambda^2 - \lambda \cos\theta (1 + S) + S = 0$, whence

$$\lambda_\pm = \tfrac{1}{2} \left((1 + S) \cos\theta \pm [(1 + S)^2 \cos^2\theta - 4S] \right). \tag{48}$$

If $S = 0$, we obtain $\lambda_+ = \cos\theta$, with eigenvector $u_+ = \binom{0}{1}$; and $\lambda_- = 0$, with eigenvector $u_- = \binom{1}{0}$. As we started from u_+, the vector $\mathbf{m}$ is shortened by a factor $\cos\tau$ at each measurement.

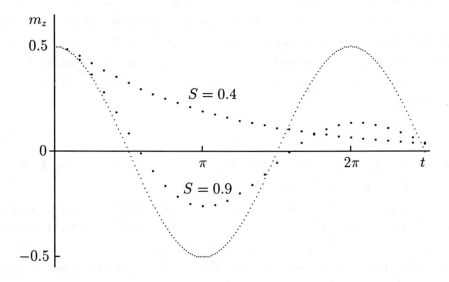

FIG. 4. Under the same conditions as in Fig. 3, each set of circles represents consecutive values of m_z that are observed if *fuzzy* measurements are performed at intervals $5\pi/60$. The coherence factors $S = 0.4$ and 0.9 correspond to $\Delta q = 0.37$ and 1.09, respectively. The critical value of S determining the onset of oscillations of m_z is $S_{cr} = 0.589$.

When $0 < S < 1$, the eigenvalues $\lambda_\pm$ may be real or complex conjugate, according to the sign of $(1 + S)^2 \cos^2\theta - 4S$. Note that this expression vanishes for $S = (1 - \sin\theta)/(1 + \sin\theta)$. Figure 4 illustrates the case $\theta = 5\pi/60$, with $S = 0.4$ and 0.9, corresponding to $\Delta q = 0.37$ and 1.09, respectively. We see that the larger Δq gives complex eigenvalues to Ω, and yields damped oscillations of m_z. Similar results were recently obtained by Milburn (1988).

The worst case scenario is a pair of consecutive measurements performed at $\tau = \pi/2$ and $\tau = \pi$, respectively. The result can easily be obtained in closed form and will later be compared with the one for a rotor of spin j. From (47) we have, with $\sin\theta = 1$,

$$\Omega^2 = \begin{pmatrix} -S & 0 \\ 0 & -S \end{pmatrix}, \tag{49}$$

so that the initial $\mathbf{m}$ is simply multiplied by $-S$, after two measurements. For very fuzzy measurements ($\Delta q \gg 1$), $-S \simeq -1$ and the spin has been flipped, as expected. For sharp measurements, $\mathbf{m}$ is reduced to zero and, thereafter, no further information is available.

Obviously, a spin $\frac{1}{2}$ particle cannot mimic the classical rotor described by Eq. (39). We therefore turn our attention to spin j. Let us prepare the rotor in an eigenstate of J_z, with $j_z = j$. Its rotation through an angle $\pi/2$ is generated by the unitary matrix U. We have, with Wigner's notations (Wigner 1959), $U_{mn} = \mathcal{D}^{(j)}(\{0\theta 0\})_{nm}$. In the "worst scenario" mentioned above, $\theta = \pi/2$ and the first measurement of J_z is performed on an eigenstate of J_x with eigenvalue j. The probability of getting $j_z = m$ is a binomial distribution

$$w_m = (2j)!/[2^j(j+m)!(j-m)!], \tag{50}$$

with variance $j/2$ (Feller 1968). This is shown in the upper diagram of Fig. 5, for the case $j = 32$ (the standard deviation then is $\Delta J_z = (j/2)^{1/2} = 4$).

Thereafter, the situation depends on whether the first measurement was sharp or fuzzy. If it was sharp ($S_{mn} = \delta_{mn}$) the histogram of expected results for the second measurement (at $\theta = \pi$) is given by the second diagram of Fig. 5: the distribution is almost uniform, and very little information is available. Better results are obtained if the first measurement is fuzzy (Busch 1987, 1988). The following diagrams of Fig. 5 show the distribution of results of the second measurement, depending on the Δq of the meter which was used for the first measurement. Obviously, a broad Δq allows the quantum state to reassemble near $j_z = -j$ (which would be its expected value at $\theta = \pi$, if the first measurement were not performed). In the particular case discussed here, I found empirically that the values of the largest components, $w_{-j}, w_{1-j}, \ldots,$ were almost independent of j (for $j > 30$) provided that Δq was scaled as $\sqrt{j}$.

On the other hand, if Δq is too broad, the "measurement" becomes useless. Actually, the expression which should be optimized is given by Eq. (11). From the example discussed above, it appears that the fuzziness of the meter should be about the same as the natural width of the detector's wave packet. A poorer resolution obviously

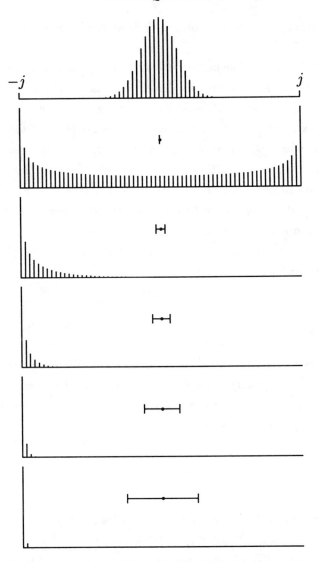

FIG. 5. The top diagram is the expected probability distribution for the results of measurements of J_z, following a preparation of the state with $j_z = +j$, and a rotation by $\pi/2$. The calculations were done for $j = 32$ and the standard deviation for this diagram is $\Delta J_z = 4$. The five other diagrams are the expected probability distributions for a subsequent measurement of J_z, after another rotation by $\pi/2$. The results depend on the resolution of the meter which performed the first measurement. From top to bottom, Δq (shown as a horizontal error bar) is 0, 1, 2, 4 and 8. (Different vertical scales were chosen in the various diagrams, for better visibility. The sum of lengths of the vertical bars is always 1, by definition.)

gives inaccurate results, but a finer resolution destroys the information in which we are interested.

This problem is peculiar to quantum systems. It disappears in the semiclassical limit, where eigenvalues become extremely dense. From the quantum point of view, *classical measurements are always fuzzy.* This is why a watched pot may boil, after all: the observer watching it is unable to resolve the energy levels of the pot. Any hypothetical device which could resolve these energy levels would also radically alter the behavior of the pot. Likewise, the mere presence of a Geiger counter does not prevent a radioactive nucleus from decaying (Peres 1987). The Geiger counter does not probe the energy levels of the nucleus (it interacts with decay products whose Hamiltonian has a continuous spectrum). As the preceding calculations show, peculiar quantum effects, such as the Zeno "paradox," occur only when individual levels are resolved (or almost resolved).

Acknowledgments

Most of this work was done during a visit to the Institute for Advanced Study, Princeton, New Jersey, whose hospitality is gratefully acknowledged. Work done at my home institution (Technion, Haifa, Israel) was supported by the Gerard Swope Fund and by the Fund for Encouragement of Research.

References

Bohm D 1951 Quantum Theory (New York: Prentice-Hall) p 120

Braunstein S L and Caves C M 1988 *Found. Phys. Lett.* **1** 3

Busch P 1987 *Symposium on the Foundations of Modern Physics. The Copenhagen Interpretation 60 Years after the Como Lecture* ed P Lahti and P Mittelstaedt (Singapore: World Scientific) p 105

Busch P 1988 *Phys. Lett. A* **130** 323

Caves C M, Thorne K S, Drever R W P, Sandberg V and Zimmerman M 1980 *Rev. Mod. Phys.* **52** 341

Caves C M 1986 *Phys. Rev. D* **33** 1643

Caves C M 1987 *Phys. Rev. D* **35** 1815

Caves C M and Milburn G J 1988 *Phys. Rev. A* **36** 5543

Chiu C B, Sudarshan E C G and Misra B 1977 *Phys. Rev.* D **16** 520

Dicke R H 1989 *Found. Phys.* **19** 385

Feller W 1968 *An Introduction to Probability Theory and its Applications* (New York: Wiley) p 228

Fleming G N 1973 *Nuovo Cim.* A **16** 232

Komar A 1962 *Phys. Rev.* **126** 365

Kraus K 1983 *States, Effects, and Operations: Fundamental Notions of Quantum Theory* (Berlin: Springer)

Lamb W E Jr 1987 *Quantum Measurement and Chaos* ed E R Pike and S Sarkar (New York: Plenum) p 183

Milburn G J 1988 *J. Opt. Soc. Am.* B **5** 1317

Misra B and Sudarshan E C G 1977 *J. Math. Phys.* **18** 756

Peres A 1980a *Phys. Rev.* D **22** 879

Peres A 1980b *Ann. Phys. (N. Y.)* **129** 33

Peres A 1980c *Am. J. Phys.* **48** 931

Peres A 1984 *Am. J. Phys.* **52** 644

Peres A 1986 *Am. J. Phys.* **54** 688

Peres A 1987 *Information Complexity and Control in Quantum Physics* ed A Blaquière, S Diner and G Lochak (Vienna: Springer) p 235

Peres A 1988 *Found. Phys.* **18** 57

Peres A and Wootters W K 1985 *Phys. Rev.* D **32** 1968

Schrödinger E 1935 *Naturwiss.* **23** 807

Singh I and Whitaker M A B 1982 *Am. J. Phys.* **50** 882

Stapp H P 1972 *Am. J. Phys.* **40** 1098

von Neumann J 1955 *Mathematical Foundations of Quantum Mechanics* (Princeton: Princeton Univ. Press) p 379

Wigner E P 1959 *Group Theory* (New York: Academic Press) p 167

Zurek W H 1984 *The Wave-Particle Dualism* ed S Diner *et al*, (Dordrecht: Reidel) p 515

No-collapse Versions of Quantum Mechanics

Yoav BEN-DOV

1. INTRODUCTION

More than sixty years after the general acceptance of Quantum Mechanics, and notwithstanding its enormous practical success, the nature of the fundamental objects appearing in its framework is unclear, and the debate concerning its conceptual foundations still goes on. This paper tries to examine what may be claimed to be one of the most 'properly quantic' (Lévy-Leblond 1977) approaches to this problem, namely 'no-collapse' versions of the non-relativistic quantum mechanics of closed systems: Everett's theory and the 'pilot wave'.

The relation between these two 'no-collapse' versions, which might appear quite disparate at first sight, was first mentioned by Bell (1976, 1981, 1987), and more recently by Zeh (1988). It is based on the fact that in both versions, all physical systems (including apparatus, cats, observers etc.) are supposed to have quantum wave functions which always develop according to the Schrödinger equation and never 'collapse'. Before discussing Bell's contention about their relation, which allows us to group them together under the same heading, we shall review separately each one of these versions, as well as their mathematical formalism and some of the problems that they entail.

2. EVERETT'S THEORY

Consider the following example. A closed system consists of a spin-1/2 quantum microobject (say, an electron), a measuring apparatus for the microobject z-spin component, and an observer reading the apparatus. Suppose that before the measurement, the microobject state is

$$\phi = \alpha|+\rangle + \beta|-\rangle, \quad \text{with} \quad |\alpha|^2 + |\beta|^2 = 1 \tag{1}$$

where $|+\rangle$ and $|-\rangle$ are the 'up' and 'down' z-spin eigenstates. In the usual formulation of quantum mechanics, ϕ is 'collapsed' during the measurement into $|+\rangle$ or $|-\rangle$ with respective probabilities $|\alpha|^2$ and $|\beta|^2$, and in accordance with the observer's reading 'up'

† This article summarizes the main themes of a Doctorate thesis prepared under the guidance of Michel Paty, Equipe REHSEIS, CNRS, France.

or 'down'. But as the 'collapse' cannot be reduced to a 'normal' Schrödinger evolution (Von Neumann 1955), this procedure implies a conceptual 'cut' in the measurement chain somewhere between the apparatus (or the observer), which is left out of the detailed description, and the microobject. Worse still, there is not even a well-defined criterion to distinguish between a measurement process, in which the 'collapse' procedure is invoked, and a 'normal' unitary evolution of the wave function. Thus, one appears to be 'cutting' the world somewhere between the observer and the observed in a vague and unprecise way (Bell 1981). This is the so-called 'measurement problem'.

In Everett's formulation of QM, there is no 'collapse'. All subsystems are treated alike, and all are assigned quantum wave functions. If the apparatus quantum state before the measurement is $|o\rangle$ and the observer state $|O\rangle$, then the composite system state before the measurement is

$$\Phi_0 = \phi|o\rangle|O\rangle. \tag{2}$$

By linearity, after the measurement and the instrument pointer reading, the composite state becomes

$$\Phi = \alpha|+\rangle|u\rangle|U\rangle + \beta|-\rangle|d\rangle|D\rangle \tag{3}$$

with $|u\rangle$ the apparatus state of having measured 'spin up', $|U\rangle$ the observer state of having read $|u\rangle$, etc. When discussing Everett's theory, it is customary to refer to superposition terms such as those appearing here as 'branches'. Thus, we have one branch in equation (2) and two in equation (3). Generally, in a similar representation, each quantum measurement multiplies the number of branches.

3. THE MANY-WORLDS INTERPRETATION

For several years following its original publication, Everett's (1957, 1973) theory went almost unnoticed. Only from around 1970 onwards, following B S DeWitt's (1968, 1970, 1971a, b) efforts to bring it into the attention of a larger public, did it start to become more widely known. But the version of Everett's theory which became most popular, and which is sometimes referred to as if it were identical with Everett's original proposal, was a somewhat simplified one, known as 'The Many-Worlds Interpretation'.

This interpretation regards the different 'branches' of the composite wave function as separate coexisting 'worlds'. As a measurement process generally creates several 'branches' out of each one that it starts with, these 'worlds' are constantly multiplying by 'splitting' from each other in a tree-like structure (although the rare possibility of 'worlds coalescence' also exists). The apparatus and the observer within each world are also assumed to split, so that each 'copy' reads a specific measurement result, valid for its own world, and shares with its other-worlds counterparts the memory record of a common experimental preparation.

The 'many-worlds' image may perhaps serve as a useful heuristic tool for a first introduction to the Everett formulation. But if taken too literally, as is sometimes the case, it leads to consequences which are not necessarily implied by the original mathematical formalism. That is, if one wishes to hold that the measurement process actually 'splits' a single world in two (or more), then the splitting 'worlds' must be regarded as distinct objects, whose number at any moment is well-defined. And as an actual multiplication of things, the 'split' itself must be considered as a real physical event, taking place instantaneously (and at some well-defined moment) at the whole spatial extension of each 'world' whenever a quantum measurement is effected. One thus arrives at a quite spectacular world-view, which has probably won for Everett's theory more publicity then credibility.

But these features are not necessarily implied by the mathematics, and indeed it is quite probable that Everett's original position was much more loyal to the mathematical formalism than is the 'many-worlds' interpretation. In fact, as Everett (1957) is clearly aware, the 'branches' (or, for him, simply 'elements of the superposition') appearing in equations like (2) and (3) are just vectors in Hilbert space. Thus, their 'number' is representation-dependent, and no objective meaning can be given to their multiplication. In a similar vein, instead of speaking about a world-multiplying 'split' taking place at some well-defined moment, Everett stresses the continuity of the measurement process, which (from the point of view of what he calls 'the complete theory') consists of nothing more than a normal unitary evolution of the composite wave function.

Everett also never hints at any worldwide 'split'. For him, the only thing which changes with measurement is the state of the apparatus (or the observer), which becomes correlated with the state of the microobject by interacting with it. It is also significant to note that in both of his published papers, Everett (1957, 1973) never uses the term 'worlds', and even the word 'split' appears in the article which he originally published (Everett 1957) only in a footnote added in proof, where he cites arguments put forward by 'some correspondents' who might have suggested that term in the first place.

A distinction between the mathematical formalism of Everett's formulation and its 'many-worlds' interpretation is not only a matter of historical interest, for it may help in answering (or, at least, clarifying) some of the objections raised against both schemes. For example, several authors (d'Espagnat 1971, Ballentine 1973, Cooper 1973) have remarked that in the 'many-worlds' scheme, the exact moment of the 'split' fails to be specified. In our example, it is not clear whether the 'split' happens when the apparatus measures the microobject or when the observer reads the apparatus. But this objection is valid only against the 'many-worlds' interpretation, in which the 'split' is considered as a real physical event which must be assigned a well-defined moment, and not against the original mathematical formulation of Everett's theory, in which the 'split' is arbitrary and representation-dependent.

Another difference between Everett's original formulation and the 'many-worlds' interpretation concerns the status of the unobserved branches. The 'many-worlds' interpretation, which considers these branches as fully existing 'worlds', has been accused of obliging us to accept an enormous metaphysical baggage, which makes this particular solution to the measurement problem quite unattractive (d'Espagnat 1971, Ballentine 1973, Bell 1981). But the equations themselves nowhere dictate that all the 'branches' should be considered as 'really existing' and on the same level. As a mathematical formalism, Everett's formulation is open at this point, and one may conceive of some way to distinguish between a single 'really existing' branch and all the other ones, without at the same time giving up the formal elegance of the formalism or re-introducing the problematic 'collapse'. Indeed, some more or less detailed 'single branch' interpretations of Everett's formulation have been proposed by several authors (Pearle 1967, Cooper 1973, Stapp 1980), and as we shall see in section 7, the 'pilot wave' formalism (at least in Bell's, or in a similar presentation) may also be considered as being directed along these lines.

The distinct 'worlds' of the 'many-worlds' interpretation may remind us of classical objects, which are also distinct and countable. Thus, the common tendency to interpret Everett's formulation in terms of many classical-like 'worlds' instead of a single quantum one (Lévy-Leblond 1977) may indicate just how difficult it is to get accustomed to genuine 'quantic' thinking. But the original Everett formulation also involves some fundamental problems, which cannot be avoided simply by rejecting the 'many-worlds' interpretation. We shall investigate two of these problems in the next two sections.

4. THE SPECIAL BASIS

Several authors (d'Espagnat 1971, Ballentine 1973, Cooper 1973, Bell 1981) have pointed out that the branch structure describing our perceived experience must correspond to a very special decomposition of the composite wave function into 'branches', e.g. that employed in writing down our equation (3). This decomposition gives on each branch an apparatus (and/or an observer) in a well-defined state of having recorded a specific measurement result. But in Everett's formulation, no fundamental distinction between 'apparatus states' and ordinary quantum states is supposed to exist. Therefore, it is not clear why this particular decomposition, and not any other one, should determine the observed branch structure. Another way to formulate this problem is to say that Everett's theory as it stands does not distinguish between the well-defined states and events that we observe at the macroscopic level (pointer positions to the left or to the right, cats alive or dead, etc) and all their quantically possible (but never observed) superpositions. How, then, can one get from it (or more generally, from the quantum formalism) an adequate account of the most obvious facts that we see all around us?

Mathematically, the existence of a special decomposition might correspond to a 'special basis' in the composite system Hilbert space (or in some adequate subspace of it), so that the observed branches be the projections of the composite state vector along its axes. But if such a 'special basis' is to be added to the theory, perhaps as an extra assumption, then it has to be specified in terms of the theory itself — that is, without using terms like 'apparatus states' or 'observer states' as if they were already given. The exact formulation of an adequate 'special basis' thus proves to be a non-trivial task.

As an answer to this problem, one may think of two possible sources of the 'special basis', that is of the 'classical' states that we observe at the macroscopic level. First, the emergence of the class of observed states might be related to the observer himself and to his internal structure (Squires 1987). For example, it might be possible that for reasons connected with the (unknown) conditions necessary for a brain–mind correlation, only a special set of brain states (and not their superpositions) can support actual states of mind. If such a set of 'special' brain states is admitted, then it can be used to determine (with Everett's (1957) 'relative state' procedure) a unique branch decomposition such that on each branch, the observer is in a well-defined state of having read a particular measurement result. As Everett further shows, the mathematical formalism itself guarantees branch consistency, so that on each branch the apparatus pointer reading corresponds to the microobject state. In this option, which may be called an 'observer basis', it is conceivable that in a given situation, two non-communicating observers might 'slice' the composite state into 'branches' in two different ways.

Many, however, might feel uneasy with such a 'subjectivist' solution to a physical problem. In addition, as we shall see in the next section, the 'observer basis' option makes the task of finding a solution to the second problem of Everett's formulation, namely that of the 'probability measure', more complicated. A more 'objectivist' option can be what we may call a 'physical basis': in this option, some physical and observer-independent criterion (possibly added to the theory as an extra assumption) is supposed to determine the observed 'special states' at the macroscopic level.

There have been some suggestions along these lines, including the 'Schmidt canonical form' of a bi-orthogonal decomposition proposed by Zeh (1973), and also arrived at by Deutsch (1985) using a criterion of impossibility of superluminal signalling (see also criticism of this proposal by Foster and Brown (1988)). Another proposal of a 'physical basis' is Bell's (1981) suggestion to give a special status to the position representation, which we shall further discuss in section 7.

5. THE PROBABILITY MEASURE

Another fundamental problem of Everett's formulation is that by dispensing with the 'collapse' procedure, it also gives up the usual probabilistic interpretation of the 'branch' coefficients α and β. In order to retrieve this interpretation and to show the quantitative agreement between his formulation and the standard quantum mechanical predictions, Everett (1957) considers a very long series of similar measurements performed on many identically prepared quantum systems. If we take the squared amplitude of each branch as its 'measure', then it can be shown that in the infinite-series limit, all the 'bad' branches (on which the quantum statistical predictions are strongly violated) have a total vanishing measure. From this mathematical result, Everett concludes that such branches are never observed. Thus, we are always sure to find ourselves on a 'good' branch and see the quantum mechanical predictions fulfilled. DeWitt (1970) cites a similar (but more detailed) argument formulated by N Graham, and which is sometimes referred to as the claim that 'the formalism yields its own interpretation'.

As d'Espagnat (1971) notes, the weak point of this argument is that it is based on the extra assumption that branches whose measure vanishes in the infinite limit practically do not exist. But Everett's argument was supposed to give meaning to the 'branch measure', not to rely on it. This, as DeWitt (1970) points out, introduces an 'element of circularity' into the argument, which was also noted by several other authors (Ballentine 1973, Deutsch 1985). Going to the more realistic case of a finite measurement series, the problem is made even sharper: here, the 'bad' branches still have a finite measure, however small, and a reason should be given as to why it is so unlikely that we end up on one of them. An exact formulation would say that in such a case, the small probability of finding ourselves on a 'bad' branch is proportional to the small branch measure. But such a relation between measure and probability, which has to be assumed in advance, is exactly what Everett's argument was supposed to give as its conclusion.

A way out of this difficulty may be provided by Stapp's (1980) suggestion to interpret the 'measure' of each branch (and not only in the infinite-series limit) as a probability for an observer to find himself on it. In a similar vein, Deutsch (1985, 1986) proposes to interpret an Everett branch not as a single 'world', but as a measurable set of non-splitting 'universes', with a uniform distribution of the observer's probability of finding himself in each single 'universe'. The measure of each such set is supposed, of course, to be proportional to the corresponding Everett branch measure.

This idea, which may be called 'branch sets', is much easier to apply to a 'physical basis' than to an 'observer basis', for the following reason. In the 'observer basis' option, the branch structure (including branch measures) is objectively given by the physics, and it is relatively easy to interpret branches which are already well-defined as measurable sets. On the other hand, in the 'observer basis' option the perceived branch structure is relative to the observer, and thus has no objective meaning. Therefore, one cannot define a class of 'branch sets' which would themselves be observer-independent.

When applied to a branch structure with finite branch measures, the 'branch sets' idea also suffers from the difficulty of explaining what exactly distinguishes between the numerous different elements of each set, which may roughly (and using Deutsch's terminology) be thought of as different 'universes' having exactly the same states of affairs in them. In contrast, if the underlying branch structure is continuous (that is, if the sum in equation (3) is replaced by an integral, on which the observed branch structure is supposed to be 'coarse-grained'), then the branch measure distribution may be replaced by a measure density function, and all the 'branch sets' elements may differ from each other on the fine-grained scale (Deutsch 1986). By these considerations, one may be led to prefer a 'physical' basis whose axes are the eigenstates of a continuous-spectrum operator. As we shall see in section 7, these features are offered by Bell's version of the 'pilot wave' theory, when interpreted from the viewpoint of Everett's formulation.

6. THE 'PILOT WAVE'

We shall now turn to consider the 'pilot wave' theory. The name 'pilot wave' itself was originally given by de Broglie (1927) to a formulation of quantum mechanics which he developed in detail only for the one-particle case. But this formulation was never meant by its author to be taken too seriously. In fact, de Broglie regarded it only as a simplified version of his much more ambitious 'double solution' approach, which aimed at re-formulating quantum mechanics in terms two different kinds of solutions for a single relativistic wave equation.

The two solutions were supposed to be, first, a 'singular solution' propagating in ordinary three-dimensional space, and second, a 'continuous solution' similar to the usual quantum wave function, which propagates (in the many-particle case) in $3N$-dimensional configuration space. By considering only the 'singular' solution as physical, and by regarding the 'continuous' solution as nothing more than a fictitious probability distribution, de Broglie hoped to avoid the notion of physical entities in configuration space, which for him was unacceptable (Ben-Dov 1989a). To this scheme, the 'pilot wave' theory which replaces the singularity of the physical solution with a 'material point' was meant to be only a provisory approximation.

Intimidated by the mathematical complexities of the 'double solution' approach and by the unfavorable reaction to his presentation (de Broglie 1928) of the simplified 'pilot wave' idea at the fifth Solvay conference, de Broglie abandoned his work on both schemes until 1952, when a 'hidden variables' idea similar to the 'pilot wave' theory was independently formulated and generalized to the many-particle case by Bohm (1952). But just like de Broglie, Bohm also did not consider the 'pilot wave' idea to be the end of the story. From 1954 onwards (Bohm and Vigier 1954, Bohm 1957) he was led to supplement it by hypothetical 'subquantum fluctuations' arising from a more fundamental physical level, at which the predictions of exact quantum mechanics would eventually fail. As for Bohm and Hiley's (1975, 1982, 1984, 1985) later work on the 'quantum potential' interpretation of quantum mechanics, it may be considered as an integration of elements taken from the 'pilot wave' theory proper, from the 'subquantum level' approach, and also from Bohm's (1980) more recent notion of 'wholeness and implicate order', strongly influenced by considerations of quantum nonseparability. These later developments will not interest us here.

As the 'pilot wave' idea was thus intermingled with other theoretical notions in both de Broglie's and Bohm's presentations, the common use of this term itself is sometimes ambiguous. We shall avoid this ambiguity by defining explicitly the mathematical formalism to be discussed here. In doing this, we stay closer to Bell's (1976, 1981, 1987, see also Ben-Dov 1987) more recent formulation of the 'pilot wave' theory, which is supposed to reproduce in all cases the exact experimental predictions of ordinary quantum mechanics. In accordance with the general interest of this article, we shall start immediately with the many-particle closed system case. The single-particle formulation follows immediately as a special case.

Suppose that our closed system consists of N quantum microobjects, including those that constitute macroscopic subsystems such as apparatus, cats, observers and so on. Let $r_1...r_N$ be the space coordinates of the N microobjects. We make the following four assumptions:

(1) With the complete system is associated a quantum wave function $\Phi(r_1...r_N, t)$, which always develops according to the Schrödinger equation and never 'collapses'.

Here, we disregard 'internal' degrees of freedom such as spin. As Bell (1981) shows, these may be accounted for by considering a multi-component Φ, although one may also think of other possibilities, for example the 'vortex' spin model suggested by Bohm *et al* (1955).

As we can see, assumption 1 is equivalent to the basic assumption of Everett's formulation. Therefore, the three assumptions to follow may be regarded as additions made by the 'pilot wave' scheme to the basic Everett framework. We shall investigate the possible justification of making these additions in the next section.

(2) In addition, with each quantum microobject is associated a well-defined 'supplementary' space coordinate $x(t)$, which gives (within experimental errors) the actually observed results of 'position measurements'.

It is convenient to introduce at this stage a single 'representative point' $X(t) = [x_1(t)...x_N(t)]$ in the $3N$-dimensional configuration space, which corresponds to the values of the N supplementary coordinates $x_1(t)...x_N(t)$ in ordinary space. The probability density and the dynamical law of the supplementary coordinates may be expressed in its terms:

(3) Supposing an adequate normalization, the probability distribution for the representative point X to be located at time t at any point R of $3N$-space is

$$Pr.[X(t) = R] = |\Phi(R,t)|^2. \tag{4}$$

(4) The i component ($i = 1...N$) of the $3N$-velocity of the representative point $X(t)$ is given by:

$$dX_i/dt = m_i^{-1}\text{Grad}_i \, \text{Im} \log \Phi(R,t) \tag{5}$$

where m_i is the mass of the i's microobject. It is easy to show that the validity of assumption 3, which guarantees that the 'pilot wave' formalism gives the same experimental results as ordinary quantum mechanics, is conserved by the dynamical equations of assumptions 1 and 4. Thus, it is necessary explicitly to make assumption 3 only at some initial moment, and it continues to hold at all later times. This was the position originally taken by de Broglie (1927) and later by Bell (1981, 1987), although Bohm (1952, 1953, also Bohm and Vigier 1954) prefers to regard assumption 3 as expressing a final equilibrium distribution, possibly brought about by the action of the 'subquantum level' fluctuations.

Given an adequate combination of initial conditions for Φ and X and assuming (Bell 1982, 1987) that all actual measurements are finally concerned with observations of positions of things like apparatus pointers and ink marks on paper, the four assumptions listed here give a completely determined theory, in which measurement results are fully specified and guaranteed exactly to reproduce the experimental predictions of ordinary quantum mechanics. It is to this theory that we shall refer as 'the pilot wave formalism'.

The main difference between our use of this term and other presentations of the 'pilot wave' theory concerns the interpretation of the 'supplementary' coordinates $x(t)$, which are usually thought of as the locations of material 'classical-like' particles, having not only positions and velocities (which are given in equation (5)), but also energy and momenta (Bohm 1952), and perhaps even charges. But as de Broglie (1930) remarked, the notion of such 'supplementary' energy and momentum leads to unnecessary complications. First, these quantities are generally not conserved, because the action of the Φ-field on the x-coordinates given in equation (5) is not counterbalanced by any reaction. One may try to restore conservation of energy by considering (as Bohm does) the Φ-field as a 'quantum potential'. But as long as one does not invoke some hypothetical 'subquantum level', with which the particle can exchange momentum, the problem of momentum non-conservation still remains. Second, any measurement of the microobject energy or momentum will give the usual quantum mechanical values, associated with the Φ-field and totally unrelated to the hypothetical 'supplementary' values. As all measurement results are already accounted for by our formalism, it seems that the only advantage gained by interpreting $x(t)$ as the location of a 'material particle' is the satisfaction of our predilection for a 'classical' metaphysics. The point to be made here is that the 'pilot wave' formalism can be regarded not only as a 'classical' theory capable of reproducing the quantum mechanical predictions, but also (and what is perhaps much more interesting) as a candidate for a truly 'quantic' theory, which does not rely on the concept of a classical-like 'particle'.

7. FROM EVERETT TO THE 'PILOT WAVE'

In this section we shall be interested in pursuing more explicitly Bell's idea concerning the relation between the two 'no-collapse' versions of quantum mechanics. The point of departure is Everett's formulation, which (in its original form) is equivalent to assumption 1 of the previous section. But as claimed in sections 4 and 5, the pure Everett formulation leaves open the two fundamental problems of the 'special basis' and of the 'probability measure'. One may try to solve them by adding some adequate extra assumptions to assumption 1, and different candidates for such supplementary assumptions have been mentioned above. We shall not attempt to evaluate them here, but only to outline the specific choices that lead from Everett's theory to the 'pilot wave' formalism, to be described in more detail elsewhere (Ben-Dov 1989b).

Start with the 'special basis' problem. Some arguments in favour of a continuous 'physical basis' were mentioned in section 5. Within this option, one may adopt the 'configuration basis' whose axes are the eigenvalues of the position operators of all the N microobjects in our system. Then, the decomposition of the wave function Φ into 'branches' looks like

$$\Phi = \int dr_1...dr_N \Phi(r_1...r_N)|r_1...r_N\rangle \tag{6}$$

where $|r_1...r_N\rangle$ is an exactly well-localized state of all the N microobjects. In this representation, it is possible to specify each 'branch', for example the one that we observe at any particular moment, by the configuration $X = x_1...x_N$ to which it corresponds. Thus, one has at any moment an 'observed configuration' in $3N$-space, which is equivalent to assumption 2 (we disregard the 'coarse-graining' of the actually observed positions, which is inconsequential here).

One still has to answer the 'probability measure' problem. A first step might simply be an explicit formulation of the branch measures' probabilistic interpretation as an extra assumption. In terms of our continuous 'configuration basis', this is equivalent to assumption 3. But a more natural foundation to this probability assumption may be provided by the 'branch set' idea described in section 5. In the continuous 'configuration basis', the class of all 'branch set' elements becomes a continuous ensemble of possible configurations, with density $|\Phi(R,t)|^2$ in $3N$-dimensional space. The single configuration $X(t)$ that we observe at each moment may be regarded as a typical member of this ensemble.

The possible configurations ensemble is thus defined at each moment separately. But it is possible to add continuity to this model by regarding the ensemble as a hypothetical 'Madelung' fluid in $3N$-dimensional configuration space, similar to the one considered by Bohm and Vigier (1954) but without any 'subquantum fluctuations'. From the Schrödinger equation can be derived a continuity equation for the flow of such a fluid of density $|\Phi(R,t)|^2$ in configuration space, with a current density given by:

$$J_i(R,t) = m_i^{-1} \text{Im}\{\Phi^*(R,t)\,\text{Grad}_i\Phi(R,t)\}. \tag{7}$$

This is just the expression for the usual 'probability current' of ordinary quantum mechanics. The observed configuration $X(t)$ may now be regarded as a typical fluid element, carried along with the flow. The components of its $3N$-dimensional velocity are:

$$dX_i/dt = J_i(X,t)/|\Phi(X,t)|^2 = m_i^{-1}\,\text{Grad}_i\,\text{Im}\,\log\Phi(X,t) \tag{8}$$

which is identical to assumption 4. Thus, the complete 'pilot wave' formalism is retrieved.

The 'pilot wave' formalism can thus be considered as the result of specific choices made within the framework of Everett's formulation — physical 'configuration basis', 'branch sets', continuous trajectories — and where, in any case, one choice or another must be made in order to overcome the specific problems discussed above. But the interpretation offered here to the 'supplementary variable' $X(t)$ differs from that of the more classically-oriented 'pilot wave' versions: instead of regarding it as representing the locations of N separate 'material particles', it is interpreted as a single 'observed configuration' in $3N$-space, to which no energy, momentum or charges should necessarily be assigned. One still needs the relation between $X(t)$ and actual observation (which is the really important element here), but this relation does not have to be based on the classical 'particle' concept. For example, $X(t)$ may be considered as a single 'real' branch out of the complete superposition. In this way, one can accept Everett's formulation without subscribing to the idea of multiple 'real worlds'.

As the 'pilot wave' formalism represents only one option within the latitude left open by Everett's original formulation, it is possible to think of other closely related, but different choices. For example, one can replace the exactly ('delta-function') localized states of the 'configuration basis' with a basis of only approximately localized ones, and so on. In this way, a complete spectrum of different 'no-collapse' versions of the non-relativistic quantum mechanics of closed systems may be offered for investigation. But first, both Everett's theory and the 'pilot wave' formalism should be thought of without the unnecessary burden of 'classical' remnants like the distinct and countable 'worlds' of the 'many-worlds' interpretation, or the 3-space 'material particles' of the earlier versions of the 'pilot wave'.

REFERENCES

Ballentine L E 1973 *Found. Phys.* **3** 229
Bell J S 1976 *Epis. Lett.* **9** 11
Bell J S 1981 *Quantum Gravity 2* eds C J Isham, R Penrose and D Sciama (Oxford: Clarendon) pp 611–37
Bell J S 1982 *Found. Phys.* **12** 989
Bell J S 1987 *Speakable and Unspeakable in Quantum Mechanics* (Cambridge: Cambridge University Press) pp 173–80
Ben-Dov Y 1987 *Fundamenta Scientiae* **8** 331
Ben-Dov Y 1989a *Annales de la Fondation Louis de Broglie* **14** 343
Ben-Dov Y 1989b *Everett's Theory and the 'Pilot Wave'* to be published
Bohm D 1952 *Phys. Rev.* **85** 166, 180
Bohm D 1953 *Phys. Rev.* **89** 458
Bohm D 1957 *Observation and Interpretation* ed S Körner (London: Butterworth) pp 33–40
Bohm D 1980 *Wholeness and the Implicate Order* (London: Routledge & Kegan Paul)
Bohm D and Hiley B J 1975 *Found. Phys.* **5** 93
Bohm D and Hiley B J 1982 *Found. Phys.* **12** 1001
Bohm D and Hiley B J 1984 *Found. Phys.* **14** 255
Bohm D and Hiley B J 1985 *Phys. Rev. Lett.* **55** 2511
Bohm D, Schiller R and Tiomno J 1955 *Suppl. al Nuovo Cimento* Serie X **1** 48
Bohm D and Vigier J-P 1954 *Phys. Rev.* **96** 208
de Broglie L 1927 *J. Phys. Rad.* Série VI **8** 225
de Broglie L 1928 *Electrons et Photons, Inst. Int. de Phys. Solvay* (Paris: Gauthier-Villars)
de Broglie L 1930 *Introduction `a l'étude de la Mécanique Ondulatoire* (Paris: Hermann)
Cooper L N 1973 *The Physicist's Conception of Nature* ed J Mehra (Dordrecht: Reidel) pp 668–83
Deutsch D 1985 *Int. J. Theor. Phys.* **24** 1

Deutsch D 1986 *The Ghost in the Atom* eds P C W Davies and J R Brown (Cambridge: Cambridge University Press) pp 83–105

DeWitt B S 1968 *Battelle Rencontres* eds C DeWitt and J A Wheeler (New York: Benjamin) pp 308–32

DeWitt B S 1970 *Phys. Today* **23** no 9 30

DeWitt B S 1971a *Proc. Int. School of Phys. 'ENRICO FERMI' Course 49* ed B d'Espagnat (New York: Academic) pp 211–62

DeWitt B S 1971b *Phys. Today* **24** no 4 41

d'Espagnat B 1971 *Conceptual Foundations of Quantum Mechanics* (New York: Benjamin)

Everett H 1957 *Rev. Mod. Phys.* **29** 454

Everett H 1973 *The Many-Worlds Interpretation of Quantum Mechanics* eds B S DeWitt and N Graham (Princeton: Princeton University Press) pp 3–140

Foster S and Brown H 1988 *Int. J. Theor. Phys.* **27** 1507

Lévy-Leblond J-M 1977 *Quantum Mechanics, a Half Century Later* eds J Leite Lopes and M Paty (Dordrecht: Reidel) pp 171–206

Von Neumann J 1955 *Mathematical Foundation of Quantum Mechanics* (Princeton: Princeton University Press)

Pearle P 1967 *Am. J. Phys.* **35** 742

Squires E J 1987 *Eur. J. Phys.* **8** 171

Stapp H P 1980 *Found. Phys.* **10** 767

Zeh H-D 1973 *Found. Phys.* **3** 109

Zeh H-D 1988 *Found. Phys.* **18** 723

An Attempt to Understand the Many-worlds Interpretation of Quantum Theory

E.J.Squires

1. THE EVERETT ASSUMPTION

The general class of what are called "Many-worlds" interpretations of quantum theory
are entirely appropriate for this book, since they all originate from the paper of Everett
(1957) which was based on the crucial assumption that the state vector of any system,
$|\Psi>$, changes with time only in accordance with the time-dependent Schrödinger
equation:

$$i\hbar \frac{\partial}{\partial t} |\Psi> = H|\Psi>$$ (1)

Thus collapse, as an additional mechanism not given by this equation, is forbidden.
The same thing is true in most "hidden-variable" versions of quantum theory, but no
such variables are postulated in Many-worlds models; here (1) is intended to be the
complete description of physics.

It is clear that the Everett assumption is very satisfactory and natural; surely one of
the messages of 20th century physics is that quantum theory is correct. The
assumption eliminates the necessity of having to introduce a "fuzzy-boundary" between a
microscopic system and a macroscopic apparatus, and it does not require the addition of
stochastic, non-linear, terms to the Schrödinger equation (e.g. Ghirardi et al., 1986 and
1989, Pearle, 1976 and 1988, and references therein). However, since the earliest days
of quantum theory there has been a widely held belief that such a simple assumption is
not compatible with the observational evidence; this is why some sort of additional
"collapse" has always been a part of orthodox quantum theory. In this article we
shall examine the reasons for this belief. We shall argue that although the difficulty
lies in a different place to that where it is sometimes located, there remains a real
problem of reconciling (1) with the totality of all experience. As a possible solution
to this problem we shall suggest a role for conscious mind as a non-physical entity.
Further study of the implications of this idea leads to speculations regarding the

experience of free—will and to the need for universal consciousness.

2. A SIMPLE MEASUREMENT

To develop these ideas we consider an explicit, simple, system.   First, we take an electron moving along the x—axis and we suppose that it has a spin—state given by

$$| \ > \ = \ \alpha | + > \ + \ \beta | - > \tag{2}$$

where $| \pm >$ are eigenstates of the z—component of spin:

$$\sigma_z | \pm > \ = \ \pm \tfrac{1}{2} | \pm > \tag{3}$$

and α, β are complex numbers.   For later use we make the totally trivial remark that the choice of this particular basis in which to expand contains no physics;   any other basis would do equally well;   the left hand side of (2) is <u>one</u> state which we have chosen to write in terms of two base vectors.

Now we allow the electron to pass through a Stern—Gerlach variable magnetic field in the z—direction.   What this means in an actual experiment is that the electron will be deflected in the "upward" ("downward") z—direction, say, according to whether its spin is + (−).   The wavefunction, however, will change in accordance with (1) which will lead to a state which we can write as

$$| \ > \ = \ \alpha | +,\text{up} > \ + \ \beta | -,\text{down} >. \tag{4}$$

Here of course the state—vector lies in a larger space than that in (2) because we are including the z—component of its momentum.   We shall continue to utilise such looseness in our notation, since no confusion should be caused.   There is a more serious possible criticism of our simplified notation, which arises from the fact that we shall ignore all but the <u>relevant</u> degrees of freedom.   For example, the apparatus that produces the magnetic field, which we have used to observe the electron spin, is a macroscopic system in thermal equilibrium with its environment.   We would not know its exact quantum state so, if this was relevant, we would have to use a statistical mixture, thereby adding the uncertainties of <u>classical</u> statistical mechanics to those of quantum theory.   In several articles in this book the view will be taken that this is an extremely important aspect of the measuring problem of quantum theory (see also Partovi, 1989).   Here, as in other articles on the Many—worlds interpretation, we shall take the contrary view, and assume that these extra complications do not help our understanding but are simply an added source of confusion.

The direction of the magnetic field in the measuring apparatus is such that the decomposition used in (2) is convenient and natural. If we had used a different basis then there would be no direct correlation between the z−component of spin and the momentum. Nevertheless we <u>could</u> use a different basis, and it is still important to note that (4) refers to <u>one</u> state (i.e. one "world", not two!).

In one sense (4) represents a perfect measurement, i.e. it is a theoretical physicist's measurement rather than one performed by an experimentalist! This is because there is <u>exact</u> correlation between spin and momentum; no terms $|-,up>$ or $|+,down>$ appear. In another, very important, sense, however, (4) does not represent a measurement at all. The key to a measurement is that it produces an answer, that is <u>one</u> answer, and (4) does not show this − it still contains both results. It is here that we see evidence of conflict between (1) and experience. An individual electron, we tend to believe, will always go either up or down when it passes through the apparatus. However, is this true? Why do we claim that it is true? The reason, the <u>only</u> reason, is that we "see" one result. Now in order for us to see an electron we need some additional detection apparatus. We could include this in the argument but, as is well known, it makes no difference, so we jump straight to the experience, or rather to the nearest thing to experience we know, which is presumably, at least for my experience, the state of my brain. Thus, allowing the oversimplification of notation noted above, the state vector becomes

$$| > = \alpha|+,up,Me^+> + \beta|-,down,Me^-> \tag{5}$$

where $|Me^{\pm}>$ corresponds to my brain having registered seeing the electron going in the up (down) direction, i.e. knowing in which direction it went. This is the prediction of physics, if physics is given by (1) and if we are able to regard "knowing" as simply a physical change in the brain. The wavefunction contains two states of Me, one that has seen up spin and one that has seen down spin. Whether we can accept this as a true view of "reality" is something that seems to be beyond the power of physics to tell us. I certainly experience one result, but it seems to be possible that this is simply because "I" only have communication with the part of the wavefunction that has experienced that result, and there is another "I", in another part of the wavefunction, which claims the opposite result.

Here we see the origin of the name "Many−worlds". Both results of the experiment, that is, my experience of up and my experience of down, are part of the physical reality which is described by the wavefunction. I can regard one experience as belonging to one world and the other as belonging to a different world; then, as a result of my observation I can say that the world has branched into two. However, it is really only because I have introduced the idea of consciousness into the discussion, that I might speak in this way. As we have emphasised, (5) is still nothing more than an expansion of <u>one</u> state vector in a complete basis, and any other basis could have been used. There is still no "physics" in the choice of a particular basis.

We now enquire into what happens when another observer (You) also measures the electron–spin. You might do this by asking me, by looking yourself at whatever detection apparatus I used, or perhaps by looking at the same computer print–out; the details are not relevant. If we include the appropriate degree of freedom of your brain, the state vector will become

$$| \ \ > \ = \ \alpha|+,up,Me^+,You^+> \ + \ \beta|-,down,Me^-,You^-> \qquad (6)$$

in an obvious notation. The correlation is again perfect; the You that has experienced one result can only interact with the Me that has experienced the same result. Thus, consistency and repeatability of experiments is ensured. It is important to realise that, in spite of what has sometimes been suggested, wavefunction collapse is not necessary for this consistency; it is a direct consequence of quantum theory as given by (1).

We should qualify the above remarks by noting that, <u>in principle</u>, there is the possibility that the two parts of the wavefunction given in (6) might interfere, so the state is not the same as the <u>mixed</u> state in which the world is described by <u>either</u> $|+,up,Me^+,You^+>$ <u>or</u> $|-,down,Me^-,You^->$. Although we know that for macroscopic systems of the size and complexity of people (or even photographic plates, etc.) interference is <u>in practice</u> impossible, because of an irreversible loss of coherence, (6) is not a mixed state and it would be a mistake to so regard it.

3.   THE ANTHROPIC ARGUMENT FOR THE MANY-WORLDS INTERPRETATION

We have seen that the Everett assumption is simple and natural, and noted that it is consistent with repeatability of observation. Here we add a positive advantage of the assumption, an advantage which makes it of particular interest to cosmologists. To do this we must first make what to some might seem to be an absurd extrapolation in scale: we go from a single electron state (2) to a wavefunction which describes the state of the universe. It is by no means obvious what sort of variables would be required for such a wavefunction and it is unlikely that it would satsify an equation like

(1), except perhaps in a suitable frame of reference. However, we shall ignore such deep issues here and note that we would certainly expect the wavefunction of the universe to be something of great complexity, such that, when expanded into a suitable basis, it would contain parts with very different properties (analogues of "up" and "down" spin values for the electron). In many current models of elementary particles, e.g. superstrings, essentially all the properties of low-energy, i.e. observable, physics depend on the nature of the vacuum. This vacuum state is actually a quantum variable, with a range of possibilities weighted by some quantum-wavefunctional. Thus essentially everything — the dimensionality of space-time, the value of the fundamental constants, even the things we regard as the basic laws of physics such as Lorentz invariance, etc. — depends upon where we look in the wavefunction. It is therefore not unreasonable to believe that in some parts of this quantum state the conditions might be right to permit the development of galaxies, of large atoms, of complex molecules, of life, even of creatures like us. That the probability for this to occur, i.e. that the relevant value of $|<\xi|\Psi>|^2$ might be infinitesimally small, is of no significance. Provided it is not zero, we exist where we are possible and that is the part of $|\Psi>$ that we observe. As noted by Tipler (1986), ideas of this sort are the ultimate limit of the process begun by Galileo; not only are we not the centre of the universe, now we see that our universe in in some sense only a tiny part of "all that is". There is nothing surprising that the universe around us seems "unlikely", we ourselves are unlikely, so we only exist in that part of the wavefunction where unlikely things happen.

4. THE PROBLEM WITH THE EVERETT ASSUMPTION

Everything in the above discussion is extremely satisfactory but, nevertheless, the model suffers from a (almost) fatal flaw. We can see this immediately in the original assumption, which we rephrase as the statement that the only change with time is that given by the deterministic Schrödinger equation — nothing else ever happens. The difficulty comes when we confront this with the well-known and experimentally established probability rule of quantum theory. Referring to (2), $|\alpha|^2$ and $|\beta|^2$ are probabilities. But they must be probabilities "of" something, i.e. of something to happen, to be, or whatever. Within the many-worlds model of quantum theory, as described above (and I believe this is consistent with the accepted version), there is nothing of which $|\alpha|^2$, or, more generally, $|<\xi|\psi>|^2$ can be the probability.

Some discussions of the Everett model try to overcome this problem by giving physical significance to the separation of the wavefunction into several terms like those on the right hand side of (4) and (5).   As a recent example, Giddings and Strominger (1988) assert "the measurements cause the universe to branch".   The first problem with this of course is that we are back with the old difficulty of the Copenhagen interpretation: what constitutes a measurement?   When does the branching occur and why?   A further difficulty can be seen when we ask what physical meaning can be given to "branching".   The word suggests that some things move apart – but what are these "things", in what space do they move and what equations describe this motion?

The most serious defect of this idea however is that it is not compatible with the probability law of quantum theory.   If both branches exist, what can it possibly mean to say that one branch exists "more" than another?   We could try to restore the probability rule by saying that on "measurement" (assuming we have discovered how to define this) the world of the simple example given in (4) above splits into *many* branches, with the first term of (4) occurring $N|\alpha|^2$ times and the second $N|\beta|^2$ times, where n is such that these numbers are integers (for a recent statement of this view see Stapp, 1989).   In my opinion this is just not believable (though I am not really sure that I know what it means).   I am therefore comforted by the fact that it cannot work if $|\alpha|^2$ is irrational.

Beginning with Everett (1957) there have been several attempts to show that the probability rule of quantum theory is unnecessary, and that it follows from the statement that if a system is in an eignestate of an observable operator then measurement will lead to the corresponding eigenvalue (the latest such attempt is due to Fahri et al. (1989)).   If this was true then our concern with understanding what the $|\alpha|^2$, etc., are probabilities of, would seem to be unnecessary.   However, all such attempts must fail.   A probability weight <u>has</u> to be introduced somehow.   The proofs depend upon considering the state consisting of N copies of states like that in (2), and then showing that in some sense this state, as N becomes large, is close to an eigenstate of the operator which measures the difference between the number of + and the number of – spins.   In actual fact the state can easily be seen to be orthogonal to all such eigenstates (Squires, 1989a).

We have therefore a real problem.   Something must "happen" if $|\alpha|^2$ is to be a probability.   It is not a new problem, and indeed it is presumably the reason why the many–worlds idea was implicitly abandoned (without ever having been so named) in the earliest days of quantum theory.   It is unreasonable to accept the good features of this idea without attempting to answer the question.   In the next section I shall suggest an answer.   It is not one which readily appeals to physicists, but I suggest it because I recognise the good features of the Everett assumption, and because I am unaware of any other answer.

5.   THE UNIQUE WORLD OF CONSCIOUS EXPERIENCE

We start with the assumption that the world of physics is described by a state vector $|\Psi>$, obeying (1) at all times.   Thus, within the world of physics there is nothing that happens for which $|<\xi|\Psi>|^2$ can represent the probability.   However, we <u>know</u> that this quantity is a probability;   at its simplest it is the probability that I become aware of one result.   Thus, "becoming aware", at least, must be something which is outside physics.   We do not need, or wish, to enter into a dualist/monist debate here.   We have <u>defined</u> what we mean by "physics" in a particular way, and anything outside that definition has to be "something else".   There does not seem to be any reason, however, why we could not if we so wished include that something else in a new definition of the word "physics".

Leaving such issues aside we return to our assumption and, to be explicit we refer again to the simple example of section 2.   What we are now saying is that when I observe the result of the experiment not only does physics change in accordance with (1), leading to the wavefunction given for example in (5), but that, <u>in addition</u>, I become aware of one result.   In other words my consciousness selects one of the two states on the right hand side of (5).   We emphasise that this selection does not change physics, i.e. the wavefunction, so the suggestion here differs from that in which consciousness somehow collapses the wavefunction (e.g. Wigner, 1962).

Having made this leap from the familiar world of physics (where at least some things seem to be understood, to the highly speculative one of conscious mind (where really nothing is understood), have we gained anything?   On the negative side we could say that we have not solved anything;   we have simply removed our problem from physics and put it into an area where everything is a problem!   However, at the very least, the idea does seem to have the merit of keeping the advantages of the Everett assumption without being dishonest.   More positively we see that in this way of understanding quantum theory it is conscious mind that ultimately is responsible for selecting the natural, preferred, basis for expansion of the wavefunction (see Deutsch, 1985, Foster and Brown, 1988, Ben-Dov, 1989, Squires, 1989b).   Thus, for example, particles, i.e. objects with definite positions, are in some sense creations of conscious mind.   Something has to create them because, as emphasised by Bell (1987, p.187), quantum theory is about waves;   it does not have any particles.

The interpretation that we have given here suggests some further speculations.   We shall discuss these below.

6.   PROBABILITY AND CONSCIOUS CHOICE

In order that we should obtain the usual probability law of quantum theory, it is necessary that the selection by conscious mind of the various experimental results is made at random, with a weight function given of course by $|<\xi|\Psi>|^2$.   This choice of weight is the obvious one, and indeed in his original paper Everett (1957) claimed that it is the only one that is consistent.   However, in many familiar (classical) situations there is the possibility of going from a random method of choosing, to a method whereby a specific determined choice is made.   The procedure of selecting a channel on a television set is an appropriate example here (Squires, 1988), because the selection of channel again does not affect "physics", which is here the incoming radiation.   If we do this selection "in the dark" we will obtain a probability distribution of channels (weighted according to the number of buttons on the set which are tuned to a given wavelength).   However we could instead first "turn on the light", and make a specific choice of channel.

Similarly, we might speculate that we could influence the result of a measurement by choosing the particular outcome that conscious mind becomes aware of.   It must be noted, however, that even if such a choice would be possible, it is not at all easy to see why we should know how to do it.   There is not any obvious evolutionary advantage in having learned how to influence whether we see an electron to have up–spin or down–spin!   Hence the fact that such an ability is not apparent, need not be interpreted as evidence against its possibility.   Equally we should be very cautious about accepting too readily the small amount of positive evidence that has been claimed for the ability of conscious choice to influence physical processes (see, for example, Radin and Nelson, 1988).

There is however one aspect of this possibility of choice that might already be apparent.   The physical mechanisms in the brain that are responsible for the actions of the body are known to occur on very small, almost microscopic scales.   It might therefore be reasonable to suppose that quantum effects are significant.   For example, and obviously grossly oversimplifying, we could imagine that the choice of whether I say "Yes" or "No" in answer to a particular question could depend upon whether a given electron is transmitted or reflected by some potential barrier.   The experience of making a free choice between either of these answers could therefore be associated with my selection of which particular outcome I choose (i.e. become aware of).

Incidentally I am not here suggesting that "free–will" and "determinism" are in any way incompatible.   I do not believe that they are.   I regard free–will as an experience, i.e. as a property of consciousness.   It is however an experience of something, and this way of looking at quantum theory seems to offer a possible clue to what that "something" is.   Related ideas (but not in the context of the Many–worlds model) have been discussed by Eccles (1986).

7.   THE NEED FOR A UNIVERSAL CONSCIOUSNESS

Most readers will have noted that our suggested answer to the problem of section 4 has introduced a new problem.   To see this we refer again to the equation (6) which shows the wavefunction when both you and I have observed the electron.   Both states of each of our brains exist in the wavefunction and are correctly correlated.   However something else has to have happened, namely we have both become aware of one result, i.e. only one of Me^+, Me^-, and similarly of You^+, You^-, is conscious.   If I now make the assumption that all (or at least most) of the people I meet are conscious, we can appreciate this new difficulty.   If Me^+ is conscious how does your consciousness know that it has to select You^+, rather than You^-?   How can we ensure consistency between separate observations by different observers?   This was not a problem in section 2 because then the two parts of the right hand side of (5) for example were not distinguished.   Now they are.   In terms of the analogy of switching on a television set, we have to be sure that all observers in fact see the same programme.

It seems clear that the only way of doing this is for there to be some sort of universal nature to consciousness.   When my consciousness has selected one result then this must imply that Consciousness has selected one result.   Thus, when you observe at a later time, then the choice has already been made (there is only one "tuner").   The wavefunction still contains parts corresponding to two (say) results, but for all but the first observer only one outcome is possible (thus, the probability rule of quantum theory does not hold for later observers).

The concept of a universal consciousness is not something which physicists readily accept, although people with different cultural backgrounds are usually very happy with it.   It appears to be incompatible with one of the simplest and most obvious properties of consciousness, namely its essential privacy.   It may open the door to strange effects like clairvoyance, etc. which we would (rightly) be reluctant to admit as possible.   Of course it is clear that, quite apart from the consistency argument given above, there is in this interpretation a need for some sort of universality of consciousness because without it the world would not have the non–local properties revealed by the Bell inequalities.   The Schrödinger equation (1), although it refers to an object which exists in configuration space, is local, so something has to introduce the required non–locality. We have kept physics local by hiding the non–locality in the idea of universal consciousness. There is evidence to suggest that Schrödinger would have approved (see, for example, Schrödinger, 1958); he would certainly have preferred such ideas to "quantum jumping".

References

Bell J S 1987 *Speakable and Unspeakable in Quantum Mechanics* (Cambridge)

Ben-Dov Y 1989 *An observer decomposition of Everett's theory* Tel-Aviv preprint, to be published in *Found. of Phys*

Deutsch D 1985 *Int. J. Theor. Phys.* **24** 1

Eccles J C 1988 *Proc. Roy. Soc.* **B227**, 411

Everett H 1957 *Rev. Mod. Phys.* **29**, 454

Fahri E, Goldstone J and Gutmann S *How probability arises in Quantum Mechanics* (MIT preprint, CTP 1699) to be published in *Annals of Physics*

Foster S and Brown H 1988 *Int. J. Theor. Phys.* **24** 1

Ghirardi G C, Rimini A and Webber T 1986 *Phys. Rev.* **D34** 470

Ghirardi G C, Pearle P and Rimini A 1989 *Markov Processes in Hilbert Space and Continuous Spontaneous Localisation of Systems of Identical Particles*, University of Trieste preprint

Giddings S and Strominger A 1988 *Nuc. Phys.* **B307** 854

Partovi M H 1989 *Phys. Lett.* **A137**, 440 and 445

Pearle P 1976 *Phys. Rev.* **D13**, 857

Pearle P 1988 *Combining Stochastic Dynamical State Vector Reduction with Spontaneous Localisation*, Trieste preprint (IC/88/99)

Radin D I and Nelson R D 1988 *Evidence for Consciousness Related Anomalies in Random Physical Systems*, report of Princeton University Department of Psychology

Schrödinger E 1958 *Mind and Matter* (reprinted by Cambridge University Press, 1967)

Squires E J 1988 *Found. Phys. Letters* **1**, 13

Squires E J 1989a *On an alleged proof of the quantum probability law* Durham preprint DTP 89/49

Squires E J 1989b *Why is position special?* Durham preprint DTP 89/43

Stapp H 1989 *Noise-induced reduction of wave packets* Berkeley preprint (LBL-26968)

Tipler F 1986 *Phys. Reports* **137**, 231

Wigner E 1962 in *The Scientist Speculates* ed. Good (Heinemann, London).

Uncertainty and Measurement

P.T. Landsberg

1. σ_x

This paper is about the sense in which the usual uncertainty relation

$$\sigma_x \sigma_p \geq \tfrac{1}{2}\hbar \tag{1}$$

can be considered to result from measurements. Here σ_x^2 is the variance of a position measurement for the x axis and σ_p^2 is the variance for the momentum component. For simplicity we confine attention to a one-dimensional box of length L and infinitely high potential walls. For quantum state n $(n=1,2,\ldots)$ it is readily shown that (Q stands for "quantum")

$$\sigma_{xQ}^2 = (L^2/12)\,(1-6/n^2\pi^2) \tag{2}$$

for a wavefunction $(2/L)^{\frac{1}{2}}\sin(n\pi x/L)$. Classically there is an equal probability for the particle to be anywhere in the box so that

$$\langle x^m \rangle = L^m/(m+1) .$$

This gives ("cl" stands for "classical")

$$\sigma_{xcl}^2 = L^2/12 . \tag{3}$$

Fig. 1 gives the result. This shows that the classical variance exceeds the quantum variance:

(a) $\quad 0.626 \leq \sigma_{xQ}/\sigma_{xcl} < 1$. $\tag{4}$

(The lower limit is reached for $n=1$). Also the standard deviation or "uncertainty" σ_x as a fraction of L is huge, as already noted by Peslak (1979):

(b) $\quad \sigma_{xcl}/L = 28.9\%$, $\quad \sigma_{xQ}/L \geq 18.1\%$. $\tag{5}$

As Peslak points out, for a   10m   box and a particle of any mass σ_{xQ} is
1.81m ! Results (a) and (b) cause difficulties for the interpretation of
σ_x as measurement uncertainty since much more accurate measurements are
made every day. The interpretation of σ_x must be simply that it is an
ensemble standard deviation. It is not directly connected with a single
measurement but as always in quantum mechanics, if the ensemble interpretation is adopted (Landsberg and Home 1987), with a large number of
repeated measurements. The Heisenberg microscope approach leads directly
to

$$\Delta x \Delta p \geq \hbar \tag{1*}$$

for single measurements, which is a variant of (1), but clearly quite
different in interpretation.

   That σ_{xcl} exceeds σ_{xQ} can be understood in terms of the completely
uniform classical probability distribution assumed here. The quantum
distribution is less uniform and hence has a larger σ_x . Nonetheless in
a usual and perhaps superficial interpretation of quantum mechanics one
might have expected

$$\sigma_{xq} > \sigma_{xcl} \backsim 0 . \tag{6}$$

This is clearly not the case.

2. σ_p

Consider the momentum next. We have for the   nth   quantum state

$$\sigma_{pQ}^2 \equiv \langle p^2 \rangle - (\langle p \rangle)^2 = \langle p^2 \rangle = 2mE_n = \hbar^2 n^2 \pi^2 / L^2 . \tag{7}$$

This uncertainty should reduce the large σ_{xQ} to something broadly in
accord with (1), which it does (see section 3). It should be relatively
smaller since we have based all our considerations on energy eigenfunctions, which are _expected_ to make σ_{pQ} small and σ_{xQ} large.

   Peslak (1979) argues for the classical case

$$\sigma_{pcl}^2 = \langle p^2 \rangle - (\langle p \rangle)^2 = \langle p^2 \rangle = 2mE \tag{8}$$

which makes $\sigma_{pcl} = \sigma_{pQ}$ provided one chooses the two energies to be
identical: $E = E_n$. However, it is also possible to regard the classical
case as corresponding to a microcanonical ensemble for which the energy
spread ΔE is given. Then Δp_{cl} is a solution of

$$(p-\Delta p_{c1})^2 = 2m(E-\Delta E) \ .$$

This yields (Landsberg 1988)

$$(\Delta p_{c1})^2 = 2mE\Theta \ , \qquad \Theta \equiv 2 - \frac{\Delta E}{E} \pm 2\left(1 - \frac{\Delta E}{E}\right)^{\frac{1}{2}} \ . \tag{9}$$

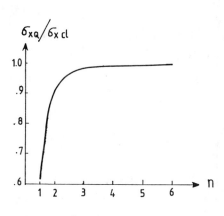

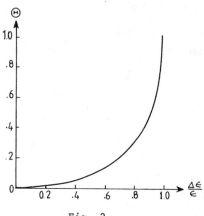

Fig. 1

σ_{xQ}/σ_{xcl} for the first few quantum numbers n .

Fig. 2

The function $\Theta \equiv \Delta p_{c1}/\sigma_{pQ}$ as a function of $\Delta E/E$.

As Fig. 2 shows, this rises from zero at $\Delta E = 0$ to unity at $\Delta E = E$, which is the case to which Peslak confined his attention. The micro-canonical interpretation has the merit that Δp can then go to zero classically, namely when ΔE is made arbitrarily small. If E is chosen to be E_n one obtains a result which is in a sense opposite to (a):

(c) $\quad 0 \leqslant \Delta p_{c1}/\sigma_{pQ} = \Theta \leqslant 1$. $\hspace{3cm}$ (10)

In order not to confuse (8) and (9) we use different symbols for the different measures of the classical "uncertainty" of p : σ_{pc1} and Δp_{c1} , the latter being a more general concept

$$\Delta p_{c1} \equiv \Delta p_{c1}\left(\frac{\Delta E}{E}\right) \ , \qquad \Delta p_{c1}(1) = \sigma_{pc1} \ . \tag{11}$$

3. $\sigma_p \sigma_x$

The quantum mechanical uncertainty product is

(d) $\sigma_{pQ}\sigma_{xQ}/\hbar = [\pi n/(12)^{\frac{1}{2}}][1-6/\pi^2n^2]^{\frac{1}{2}} \geq \left(\frac{\pi^2}{12} - \frac{1}{2}\right)^{\frac{1}{2}} = 0.568$.   (12)

Its asymptote for large n is simply $[\pi/(12)^{\frac{1}{2}}]n$ but for low n it
departs from the asymptote and reaches its smallest value at n = 1
(Fig. 3). From (12) the general principle of quantum mechanics

$\sigma_{pQ}\sigma_{xQ}/\hbar \geq \frac{1}{2}$ (13)

is satisfied. Conversely, using (12), *together with (13)*, one can obtain
an inequality for π^2 from quantum mechanics! It is

$\pi^2 \geq 9/n^2 \geq 9$.   (14)

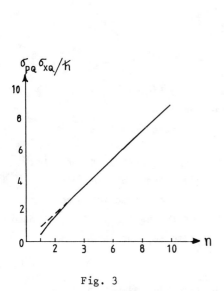

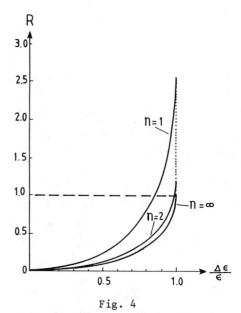

Fig. 3   Fig. 4

$\sigma_{pQ}\sigma_{xQ}/\hbar$ as a function of n .   The ratio R of the squares of
The dashes denote the asymptote.   the classical to the quantum
                                       uncertainty product as a function
                                       of $\Delta E/E$.

Following these comparable features of the quantum and classical uncertainties, one must ask for the classical uncertainty product. It is
for $E = E_n$ given by

(e) $(\Delta p_{c1} \sigma_{xc1})^2 = \dfrac{mE_n L^2}{6} \ominus = \dfrac{\hbar^2 n^2 \pi^2}{12} \ominus \begin{cases} \to 0 & (\Delta E = 0) \\ \searrow \hbar^2 n^2 \pi^2/12 & (\Delta E = E_n) \end{cases}$ (15)

Thus the use of the general Δp_{c1} leaves open the possibility of a vanishing classical uncertainty product, namely when the energy range for the microcanonical ensemble goes to zero. This seems to be a satisfactory feature of the passage from σ_{pc1} to Δp_{c1} .

Which uncertainty product is bigger? It clearly all depends on the ratio $\Delta E/E$ which we adopt (using $E = E_n$) , as shown in Fig. 4. Above the dashed line the classical product surprisingly exceeds the quantum one and this is so for all n if one chooses $\Delta E = E_n$. If this inter-pretation is adopted, only the dotted line is relevant. For $n = \infty$ Fig. 4 goes over into Fig. 2. Below the dashed line the expected situa-tion occurs and the classical uncertainty product is smaller.

It is not essential to the argument, but we note that the correlation coefficient can always be introduced into the uncertainty relation (Levy-Leblond 1986). It vanishes for our quantum mechanical case.

4. H(q)

The vast literature on the uncertainty principle continues to grow and contains additional criticisms, notably that the general uncertainty principle

$$\sigma_A \sigma_B \geqslant \frac{1}{2} |<[A,B]_\psi|$$ (16)

has a right-hand side which depends on the state ψ considered, and indeed vanishes if ψ is an eigenstate of A or B . The inequality then states merely the purely mathematical identity $\sigma_A \sigma_B \geqslant 0$ which contains no additional physics. (It is <u>not</u> a statement that an un-certainty is zero.) Inequality (1) is not subject to this criticism, but it fails on the interpretational grounds explained in sections 1 - 3. (Relation (1*) has a very different status and is not criticised here.) One could of course choose different wavefunctions for our problem and check through the argument again. Or one could consider a different problem. Still, one example seems to us to be enough to put into question the conventional interpretation of (1) (but not of (1*)) and therefore of (16) as directly relevant to measurement. Ours is an interpretational and not "merely" a mathematical criticism. In the present section we show that possible replacements are at hand, but we cannot take this matter to a satisfactory conclusion at present.

I like the suggestion of Maassen and Uffink (1988) and I give it here a simplified exposition. Let $|<a_k|\psi>|^2$, $|<b_j|\psi>|^2$ be two probability distributions which are normalised to the same sum (which may be unity). Let c be the maximum of the values $|<b_j|a_k>|$. Note that c is independent of ψ. Now it is usual in the theory of inequalities and in the theory of averages to introduce weighted means of order r, typically (Hardy et al. p.13)

$$M_r(x,q) \equiv (\Sigma q_j x_j^r)^{1/r} . \tag{17}$$

These are averages of the x_j weighted by normalised probabilities q_j. Of course, if the x_j are themselves probabilities one can choose $q_j = x_j$ and finds

$$M_r(q) \equiv (\Sigma q_j^{r+1})^{1/r}$$

and therefore

$$M_r(|<a_j|\psi>|^2) = \left\{ \Sigma_j |<a_j|\psi>|^{2(r+1)} \right\}^{1/r} . \tag{18}$$

We shall adopt the interpretation (18). Already (17) has the useful property (use l'Hopital's rule on $\ln M_r(x,p)$!)

$$\lim_{r\to 0} [\ln M_r(x,q)] = \sum_j q_j \ln x_j$$

so that $M_0(x,q) = G(x) \equiv x_1^{q_1} x_2^{q_2} \dots$

is the weighted geometrical mean. In our case $x_j = q_j$ whence

$$M_0(q) = \exp[-H(q)] , \quad H(q) \equiv -\sum_j q_j \ln q_j .$$

One can now show that our probability distributions satisfy

$$M_r(|<b_j|\psi>|^2) M_s(|<a_k|\psi>|^2) \leq c^2 \tag{19}$$

for $-\frac{1}{2} \leq s \leq 0$ and $0 \leq r = -\frac{s}{s+1} \leq \infty$. Thus in particular

$$H(|<a_k|\psi>|^2) + H(|<b_j|\psi>|^2) \geq -2 \ln c . \tag{20}$$

This is an "entropic" uncertainty relation in that uncertainty is now measured by the entropy $H(q)$. Its right-hand side is independent of ψ , and this is an advantage. We will report elsewhere on how it fares when tested on our simple model.

CONCLUSION

Using energy eigenfunctions for a particle in a one-dimensional box of side L , one would expect the "uncertainty" σ_{pQ} in momentum to be relatively small, and σ_{xQ} to be relatively large while their product satisfies the uncertainty relation. This has been verified. However, $\sigma_{xQ}/L \geq 18.1\%$ is unexpectedly large and a classical analogue is at 28.9% even larger, rather than smaller as one might have expected. This encourages one to make a sharp distinction between the uncertainties obtained by scrutinising experiments (as with Heisenberg's microscope) and the standard deviation type of uncertainty definition of the quantum mechanical formalism, which is the only one criticised here. The ensemble interpretation of quantum mechanics calls indeed for some uncertainty definition based on statistics, but a better one than that conventionally used seems required. Our main quantitative results are labelled (a) to (e).

References

Hardy G H, Littlewood J E and Polya G 1934 Inequalities (Cambridge
    University Press).
Hilgevoord J and Uffink J B 1985 Europ. J. Phys. 6 165
Landsberg P T and Home D 1987 Am. J. Phys. 55 226
Landsberg P T 1988 Foundations of Physics 18 969
Levy-Leblond J-M 1986 Am. J. Phys. 54 135
Maassen H and Uffink J B M 1988 Phys. Rev. Lett. 60 1103
Peslak Jr J 1979 Am. J. Phys. 47 39